T0140911

An investigation of microRNA target regulation mechanisms using an integrative approach

Dissertation

zur

Erlangung des akademischen Grades

Doktor-Ingenieur (Dr.-Ing.)

Promotionsgebiet Bioinformatik

Fakultät für Informatik und Elektrotechnik

Universität Rostock

vorgelegt von

Ulf Schmitz

geboren am 08.Juli 1978 in Wismar, Wohnhaft in Papendorf

Rostock, 06.Januar 2015

Dissertation der Universität Rostock

Gutachter:
Prof. Olaf Wolkenhauer (Universität Rostock)
Prof. Gero Wedemann (Fachhochschule Stralsund)
Prof. André Franke (Christian-Albrechts-Universität zu Kiel)

Datum der Einreichung: 06.01.2015
Datum der Verteidigung: 19.05.2015

Bibliografische Information der Deutschen Nationalbibliothek

Die Deutsche Nationalbibliothek verzeichnet diese Publikation in der Deutschen Nationalbibliografie; detaillierte bibliografische Daten sind im Internet über http://dnb.d-nb.de abrufbar.

ISBN 978-3-8325-4005-0

Logos Verlag Berlin GmbH
Comeniushof, Gubener Str. 47,
10243 Berlin
Tel.: +49 (0)30 42 85 10 90
Fax: +49 (0)30 42 85 10 92
INTERNET: http://www.logos-verlag.de

Abstract

MicroRNAs (miRNAs) regulate gene expression at the post-transcriptional level in most cell-biological processes. Deregulated miRNA expression can mediate the emergence and progression of human diseases. It has been shown that miRNAs can repress translation of mutual target genes in a concerted way. Thereby, a fine-tuned and adaptive target gene regulation can be achieved.

However, until now it has been unclear (i) which mechanisms underlie a collective target regulation and (ii) which miRNAs may even cooperate. After a comparative analysis of methods in computational miRNA biology this thesis describes two workflows that were designed to answer these two questions. In particular, the first workflow investigates mechanisms of collective target repression by miRNAs, while the second workflow is used to identify and analyse targets of synergistic miRNA regulation. Both workflows integrate bioinformatics and systems biology approaches with the goal for a maximum level of confidence with *in silico* predictions. More specifically, methods for data integration, the prediction of molecular interactions, modelling molecular complex structures, network analysis, and kinetic modelling are used here for the analysis of collective target regulation by miRNAs and for the prediction of cooperating miRNA pairs and their mutual target genes.

Two case studies are presented to demonstrate the efficiency of both workflows. In the first case study, the complex regulatory network of a miRNA target hub gene is analysed and predictive simulations are performed. These reveal a distinctive and fine-tuned regulation of the target hub gene for different cellular conditions. In the second case study, the human genome is analysed to predict targets of synergistic miRNA-mediated repression. The results suggest that the phenomenon of cooperative target regulation by miRNA pairs is a common cellular mechanism in humans which may be important in the context of cancer pathways.

The proposed workflows provide comprehensive tools for investigating mechanisms of post-transcriptional gene regulation and for the prediction of regulatory RNA interactions. These workflows can also be applied in the investigation of other gene regulatory networks and for predicting targets of cooperative miRNA regulation in other animal species.

Zusammenfassung

MicroRNAs (miRNAs) regulieren die Genexpression auf post-transkriptioneller Ebene in den meisten der zellbiologischen Prozesse. Deregulierte miRNA-Expression kann zur Entstehung und zum Fortschreiten von Krankheiten im Menschen führen. Es wurde gezeigt, dass miRNAs die Translation gemeinsamer Zielgene in konzertierter Weise inhibieren können. Dadurch kann eine fein abgestimmte und adaptive Zielgenregulierung erreicht werden.

Bis zuletzt war jedoch nicht klar, (i) welche Mechanismen einer gemeinsamen Zielgenregulierung zugrunde liegen und (ii) welche miRNAs dabei kooperieren können. Nach einer vergleichenden Analyse von Methoden in der computergestützten miRNA Biologie, beschreibt diese Arbeit zwei Workflows, die entwickelt wurden, um diese offenen Fragen zu beantworten. Der erste Workflow zielt darauf ab, Mechanismen der kollektiven Zielgenrepression durch miRNAs zu untersuchen, während der zweite Workflow für die Identifizierung und Analyse von Zielen synergetischer miRNA-Regulierung verwendet wird. Beide Workflows integrieren Methoden der Bioinformatik sowie der Systembiologie, um ein Höchstmaß an Verlässlichkeit mit *in silico* Vorhersagen zu erreichen. Insbesondere werden hier Verfahren zur Datenintegration, der Vorhersage von molekularen Wechselwirkungen, der Modellierung von Molekülkomplexstrukturen, der Netzwerkanalyse und der kinetischen Modellierung für die Analyse von kollektiver Zielgenregulierung durch miRNAs, als auch für die Vorhersage von kooperierenden miRNA-Paaren und ihrer gemeinsamen Zielgene genutzt.

Zwei Fallstudien werden vorgestellt, die die Effizienz beider Workflows demonstrieren. In der ersten Fallstudie wird das komplexe regulatorische Netzwerk eines „miRNA-Targethub“-Gens analysiert sowie prädiktive Simulationen durchgeführt. Diese weisen auf eine individuelle und fein abgestimmte Regulation der „Targethubs“ für unterschiedliche zelluläre Bedingungen hin. In der zweiten Fallstudie wird das menschliche Genom analysiert, um Ziele synergetischer miRNA-Repression zu identifizieren. Die Ergebnisse deuten darauf hin, dass das Phänomen der kooperativen Zielgenregulierung durch miRNA Paare einen verbreiteten zellulären Mechanismus im Menschen darstellt und im Zusammenhang mit Signalpfadwegen in Krebs wichtig sein kann.

Die vorgestellten Workflows bieten umfassende Möglichkeiten für die Untersuchung von Mechanismen der post-transkriptionellen Genregulation und für die Vorhersage von regulatorischen RNA-Wechselwirkungen. Diese Workflows können z.B. auch für die Untersuchung anderer genregulatorischer Netzwerke und für die Vorhersage von Zielegenen kooperativer miRNA-Regulation in anderen Spezies verwendet werden.

Acknowledgements

First of all I would to thank **Olaf Wolkenhauer** for giving me the opportunity to do my PhD under his supervision. Although I started as a systems engineer in the Department of Systems Biology & Bioinformatics, he encouraged and supported me towards an engagement with the scientific concerns in our department. Ever since I started working here I have enjoyed my work and appreciated the opportunities that he gave to me.

This brings me to my colleagues, many of whom I share friendship with. Many people have come and gone over the years, but I have fond memories of everyone and I'm thankful that our ways have crossed and that we have shared some of our lifetime together.

I am especially grateful to the small group of colleagues who share my interest in miRNAs. In particular **Julio Vera**, **Xin Lai** and **Shailendra Gupta** worked with me relentlessly on our visions and hypotheses. We enjoyed long and fruitful discussions without any sign of competition.

In this context I also want to thank my experimental collaborators **Manfred Kunz**, **Animesh Bhattacharya** from the University of Leipzig and the colleagues from the Institute of Experimental Gene Therapy and Cancer Research lead by **Brigitte Pützer**. I am also looking forward to continuing working with **Michael Linnebacher** and **Robert David** with whom we have already started promising collaborations.

A big "Thank you!" goes to my colleagues and friends **Sonja Strunz**, **Ulf Liebal** and **Xin Lai** for reading my thesis and giving me very valuable suggestions.

I would like to give special thanks to my friends **Jan Wetzel**, **Marko Muczinski** and **Daniel Witte**. In the last years you became an essential part of my life. It is the greatest pleasure to see you once a week to chat, drink a beer and of course perform some music together. Similarly, I want to thank **Mario Herrmann** for his everlasting friendship despite the greatest possible distances.

I am particularly grateful to my parents **Sybille** and **Uwe Schmitz.** I owe so much to you, because you have always supported me in all stages of my life. You are the biggest constant in my life and I am grateful for your love and trust. Likewise, I want to thank my sister **Ulrike Schmitz-Schlüter** and her family (**Olaf**, **Anton** and **Martha**) for their support and friendship.

My biggest thanks go to my family who supported me so much and helped to find the right work-life-balance. **Sandra**, **Carla** and **Edda** you show me every day what is really important in life.

Contents

List of Figures

List of Tables

Abbreviations

3D	three dimensional
a.k.a.	also known as
ADAR[1]	adenosine deaminases acting on RNA
AGO	Argonaute
BLAST	Basic Local Alignment Search Tool
cDNA	complementary DNA
charmm	Chemistry at HARvard Macromolecular Mechanics
CLASH	crosslinking, ligation, and sequencing of hybrids
CT	canonical triplex
DGCR8	DiGeorge syndrome critical region gene 8
ECDF	empirical cumulative distribution function
FLTR	from left to right
GO	Gene Ontology
HEK	human embryonic kidney
kcal/mol	kilocalorie per mole
lncRNA	long non-coding RNA
MD	triplex involving miRNA-miRNA hybridization (miRNA-duplex)
MDS	molecular dynamics simulations
MFE	minimum free energy
miRNA	microRNA
mRNA	messenger RNA
MSC	triplex with miRNA self-complementarity
ncRNA	non-coding RNA
nM	nano Molar
nt	nucleotide(s)
ODE	ordinary differential equation
ORF	open reading frame
P-bodies	processing bodies
PCR	polymerase chain reaction
PDB	Protein Data Bank
PDF	Portable Document Format
PE	potential energy
PMID	PubMed identifier

[1] Note, that in the manuscript gene and mRNA names are italicised according to the guidelines for human gene nomenclature (Wain *et al.*, 2002). Protein designations in contrast are not italicised.

Pol II	RNA polymerase II
PPI	protein-protein-interaction
pre-miRNA	precursor microRNA
pri-miRNA	primary microRNA
ps	pico seconds
pSILAC	pulsed stable isotope labeling by amino acids in cell culture
qRT-PCR	quantitative real time PCR
RBP	RNA binding protein
REST	representational state transfer
RG	repression gain
ribozyme	ribonucleic acid enzyme
RISC	RNA induced silencing complex
RLC	RISC loading complex
RNA	ribonucleic acid
RNAi	RNA interference
RNase III	nuclear ribonuclease III
RNA-seq	RNA sequencing
rRNA	ribosomal RNA
SBGN	Systems Biology Graphical Notation
SBML	Systems Biology Markup Language
shRNA	small hairpin RNA
siRNA	small interfering RNA
SNP	single nucleotide polymorphism
ST	stability time
TEC	triplex equilibrium concentration
TF	transcription factor
TFE	triplex free energy
tRNA	transfer RNA
TSC	triplex with target self-complementarity
URL	Uniform Resource Locator
UTR	untranslated region

1 Introduction

This chapter is based on the following own publications:

Raasch P, **Schmitz U**, Patenge N, Vera J, Kreikemeyer B, Wolkenhauer O (2010) Non-coding RNA detection methods combined to improve usability, reproducibility and precision. BMC Bioinformatics, 11:491

Freiesleben S, **Schmitz U**, Vera J (2013) MicroRNA Biogenesis, Regulation. In Encyclopedia of Systems Biology (ed. Dubitzky W, Wolkenhauer O, Cho K, Yokota H), pp. 1306-1310. Springer New York.

Schmitz U (2013) microRNA Target Regulation. In Encyclopedia of Systems Biology (ed. Dubitzky W, Wolkenhauer O, Cho K, Yokota H), pp. 1346-1350. Springer New York.

Schmitz U, Gupta S (2013) Target Site. In Encyclopedia of Systems Biology (ed. Dubitzky W, Wolkenhauer O, Cho K, Yokota H), pp. 2142-2145. Springer New York.

Schmitz U, Vearasilp K (2013). MicroRNA Web Resources. In Encyclopedia of Systems Biology (ed. Dubitzky W, Wolkenhauer O, Cho K, Yokota H), pp. 1335-1337. Springer New York.

Synopsis

This chapter traverses the background information available for the biological aspect of this thesis. However, I first motivate my work, discuss open issues in microRNA (miRNA) research that have been targeted in this thesis and promote the idea of an integrated bioinformatics and systems biology approach for the effective exploration of miRNA target regulation. Thereafter, I provide an introduction to the domain of non-coding ribonucleic acid (ncRNA) molecules, with a special focus on the class of miRNAs.

1.1 Motivation

It is beyond doubt that studying the role and functioning of miRNAs is of great importance for understanding mechanisms of gene regulation. MiRNAs are versatile, influential regulators of gene expression at the post-transcriptional level and have been shown to impact upon the emergence and progression of many diseases. Although a wealth of knowledge about miRNA-target regulation already exist, the details of some regulatory mechanisms remain obscure. For example, a target gene being regulated by two or more miRNAs or the involvement of miRNAs in regulatory networks that include transcriptional, post-transcriptional and post-translational regulation have yet to be understood. The goal of the presented work was to answer two specific questions about miRNA target regulation mechanisms:

(1) Which mechanisms underlie a collective miRNA target regulation?
(2) Which miRNAs can cooperate in the regulation of a mutual target?

This thesis is to demonstrate how to integrate tools and methodologies from bioinformatics, structural biology and systems biology to overcome shortcomings in the analysis of miRNA target regulation and to answer these two questions. In particular, ways have to be found for the reconstruction and analysis of regulatory networks of miRNA target hub genes. Furthermore, it is the aim to reliably predict and analyse targets of cooperative miRNA regulation. In summary, this thesis motivates an integrative approache in order to comprehensively address these open challenges.

1.2 Outline of the thesis

The present work is an investigation into mechanisms by which miRNAs modulate post-transcriptional regulation of target mRNAs. To help uncovering and understanding these mechanisms we have developed two integrative workflows, combining bioinformatics and systems biology approaches:

1. Integrative workflow for the re**C**onstruction and **A**nalysis of networks regulating miRNA **T**arget **H**ubs (**CATH**)
2. Integrative workflow for the **Pre**diction and validation of **Co**operating **M**iRNAs and their mutual targets (**PreCoM**)

The first chapter of the thesis introduces the biological focus on miRNAs and identifies the research gap addressed by this work.

The second chapter provides a comparative analysis of bioinformatics and systems biology approaches used in computational miRNA biology. It becomes apparent that to the present date these two domains evolve in parallel. To answer certain biological or biomedical questions it is, however, necessary to identify ways to combine these approaches in integrative workflows.

In the third chapter, the focus lies on miRNA target regulation and more specifically on the special phenomenon of collective regulation of miRNA target hubs. In this study, the regulatory network composed of transcription factors (TF), miRNAs and protein-protein interactions involved in the regulation of the miRNA target hub *CDKN1A* was reconstructed. This network is based on data derived from the literature, public databases and computational predictions. We implemented a confidence scoring system to indicate how reliable molecular interactions within this network are. Furthermore, we used a network analysis approach to identify structural and topological properties of the network. Thereby, we found an enrichment of process specific regulatory motifs, which suggests, that TFs and miRNAs realize a coordinated regulation of target hub genes. Subsequently, we derived a mathematical model composed of ordinary differential equations (ODE) which was used to predict the dynamics of *CDKN1A* expression. Our simulations demonstrate that miRNA target hubs are subject to fine-tuned transient and long-term regulation that is adjusted according to the needs in different cellular scenarios. Another result of this study is the CATH workflow which we propose for the reconstruction and analysis of regulatory miRNA target hub networks .

In the fourth chapter, the focus lies on the recently discovered mechanism of cooperative target gene regulation by pairs of miRNAs. We developed a workflow (PreCoM) for the prediction and analysis of target genes and miRNA pairs involved in this phenomenon. A wide range of bioinformatics and systems biology approaches are integrated in this workflow. We used a target prediction method which is based on a modified Smith-Waterman algorithm and position specific empirically defined rules for the identification of conserved target sites that form stable structures when hybridizing with a miRNA. Using this method and a previously observed prerequisite for miRNA cooperativity we found a wealth of putative targets of cooperative miRNA regulation. We implemented RNA structure modelling and molecular dynamics simulations (MDS) to predict the thermodynamic stability of putative triplexes formed by two cooperating miRNAs and their mutual target mRNA. Using statistical analyses we found that canonical triplex structures with preserved seed binding are typically more stable than other structural conformations. We used a partition function algorithm to compute the equilibrium probability distribution of complexes (duplexes, triplexes etc.) and monomers that can emerge from an RNA triple including two miRNAs and their mutual target mRNA. Furthermore, we derived ODE-based mathematical models for cooperative target regulation. Simulations show different landscapes of target repression which largely depend on the energy parameter values. The proposed workflow was implemented and used to analyse the human genome, which identified several thousand targets of cooperative gene regulation. The results were accumulated in a database called TriplexRNA, which supports scientists in the identification of novel targets of synergistic miRNA regulation and their induced target repression efficiency.

I conclude my work in the fifth chapter and discuss how the developed workflows can be of use in other studies. I then describe how the results broaden our understanding of miRNA-mediated regulatory mechanisms and possible consequences in a larger framework. Finally, I preview possible future directions and open questions that could be addressed with a similar computational biology approach.

Of note, in the following chapters I use the pronoun 'we' whenever I describe or discuss our own work as most of the work presented in this thesis emerged from collaborative efforts. However, I provide a detailed listing of my own contributions to the key publications in Appendix A. Moreover, at the beginning of each chapter, the publications on which it is based are indicated. A full list of publications can be found in the CV as part of the Appendix. For the purpose of the present PhD thesis the focus is on the following selected key publications:

1. X Lai*, **U Schmitz***, S Gupta, A Bhattacharya, M Kunz, O Wolkenhauer, J Vera (2012) Computational analysis of target hub gene repression regulated by multiple and cooperative miRNAs. *Nucleic Acid Research*, 40: 8818-8834, (* eq. contr. authors).

2. **U Schmitz**, X Lai, F Winter, O Wolkenhauer, J Vera, S Gupta (2014) Cooperative gene regulation by microRNA pairs and their identification using a computational workflow. *Nucleic Acid Research*, 42, 12, p. 7539-7552.

1.3 Non-coding RNA genes

There are two conceptual roles of RNA molecules. One is linked to messenger RNAs (mRNA) which are blueprints of genetic code and are used to carry the construction plan for proteins from the nucleus to the cytoplasm, where mRNA is translated by the ribosome (the translation machinery) into amino acid chains (the building blocks of proteins). However, there exist a variety of RNA molecules that play a more active role in the cell[2]. These functional RNAs are non-coding molecules (a.k.a. ncRNAs), i.e. they do not encode proteins. A stable structural fold is the basis for their biological function (Gardner and Giegerich, 2004). Most prominent representatives of this domain are ribosomal RNAs (rRNA), which are involved in protein synthesis and form a part of the ribosome, and transfer RNAs (tRNA), known as carrier of amino acids, used by the ribosome in the translation of ribonucleic acid messages (mRNAs) into peptides composed of amino acids. The discovery of another class of functional RNAs, the catalytic RNAs (a.k.a. ribonucleic acid enzymes or ribozymes) that have been found to be involved in RNA processing reactions, such as RNA splicing, viral replication, and transfer RNA biosynthesis, stimulated the discussion about the RNA world hypothesis, which suggests RNA based organisms as an essential step in evolution (Gilbert, 1986; Jeffares *et al.*, 1998; Poole *et al.*, 1998). It was then evident that the RNA molecule domain offers many more functionalities than just the conveyance of the genetic message. NcRNAs are between 20nt and 400nt in length and their genes were long believed to reside exclusively in intergenic regions, but recent studies have shown that they can also be located in open reading frames (ORFs), operons or even introns of other genes (Dinger

[2] Apart from intra-cellular RNAs, there are also RNAs that can be found outside cells, e.g. in body fluids (Chen *et al.*, 2008).

et al., 2008; Raasch *et al.*, 2010). A classification scheme for ncRNA molecules that includes the classes introduced above is provided in Figure 1.1 and a list of all ncRNA classes with a short description is provided in Table 1.1.

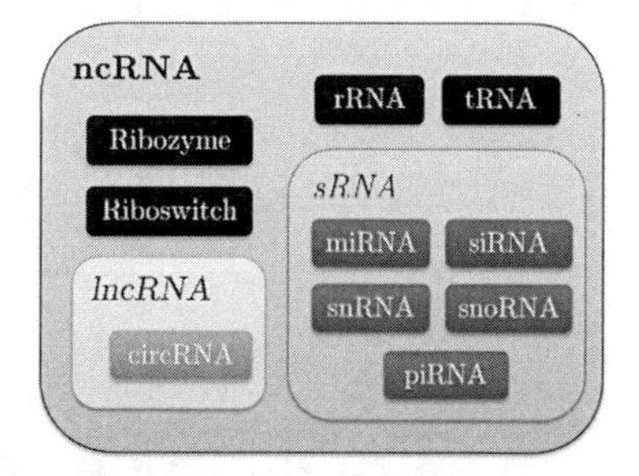

Figure 1.1 NcRNA classification scheme.
The scheme is composed of three boxes, each representing a ncRNA domain containing several classes or subdomains. In the case of lncRNAs, more classes can be expected to be found and characterized in the near future. See also Table 1.1.

Class	Description
ncRNA - non-coding RNA	all RNAs not encoding a protein
fRNA - functional RNA	synonym for non-coding RNA
rRNA - ribosomal RNA	RNA component of the ribosome
tRNA – transfer RNA	amino acid carrier molecule
Ribozyme - ribonucleic acid enzyme	involved in RNA processing reactions
Riboswitch	regulatory segment of a messenger RNA
siRNA - small interfering RNA	active molecules in RNA interference
snRNA - small nuclear RNA (a.k.a. U-RNA)	include spliceosomal RNA and snoRNA
snoRNA - small nucleolar RNA	mediate chemical modifications of e.g. rRNAs, tRNAs and snRNAs
piRNA – piwi-interacting RNA	small RNAs interacting with piwi proteins in germ line cells
miRNA – microRNA	post-transcriptional regulators of gene expression
lncRNA – long non-coding RNA	non-coding transcripts composed of more than 200 nucleotides
circRNA – circular RNA	covalently closed continuous RNA loop (can act as miRNA sponge)

Table 1.1 Different classes of ncRNAs.

The experimental identification of novel ncRNA genes requires prior knowledge about the underlying genome (i.e. full or partial genome sequence) and is restricted to those ncRNAs that are expressed at the time of the experiment. Frequently used techniques are: direct sequencing of ncRNAs, shotgun cloning of small-sized ncRNAs (cDNA libraries), microarrays and the recently invented whole transcriptome shotgun sequencing a.k.a. RNAseq (RNA sequencing; Hüttenhofer and Vogel, 2006). However, these approaches have their limits when performed without prior bioinformatics predictions. Therefore, sophisticated algorithms are required that predict ncRNA genes with high specificity to save time and experimental misfortune. In our own previous work we developed a software framework named *moses* (Raasch *et al.*, 2010), that integrates established methods for a more sensitive and accurate prediction of novel ncRNA genes. We used our software to identify novel ncRNA genes in the gram positive bacterium *Streptococcus pyogenes*. Experimental validation confirmed our predictions and lead to the discovery of four previously unknown ncRNA genes (Raasch *et al.*, 2010).

1.4 MiRNAs

Another recently discovered representative of the ncRNA domain and focus of this thesis is the miRNA, an abundant class of short single stranded RNA molecules that can regulate the expression of genes at the post-transcriptional level (Lau *et al.*, 2001). Their potential to regulate the expression of a large portion of the human genome makes miRNAs a versatile tool for a context-specific regulation of most cellular processes. Deregulation of miRNAs, however, can lead to the emergence and also the progression of human pathologies, including cancer (Soifer *et al.*, 2007). Therefore, this class of ncRNAs forms an integral part of present and future biochemical, cell biological, biomedical and clinical investigations (Kozomara and Griffiths-Jones, 2011; Landgraf *et al.*, 2007; Selbach *et al.*, 2008; Garzon *et al.*, 2009).

With the discovery of post-transcriptional gene regulation by RNA interference mechanisms (e.g. miRNA regulation), the long-time established central dogma of molecular biology, which describes genes expression as a linear process, had to be revised (Figure 1.2).

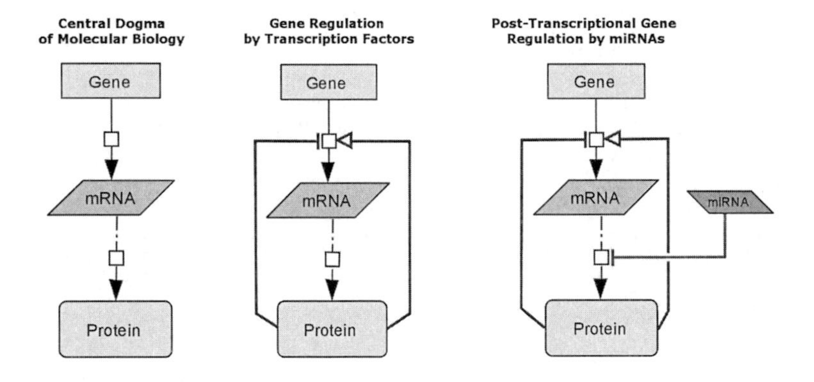

Figure 1.2 Evolution of the central dogma of molecular biology.
Our view on gene expression has evolved from the assumption of a simple and linear sequence of events to a complex and sophisticated regulatory structure. An important piece in this puzzle is the recently discovered miRNA which can regulate gene expression at the post-transcriptional level.

The first miRNA, lin-4, was discovered in 1993 in the nematode *Caenorhabditis elegans* and was found to regulate the protein synthesis of lin-14 which plays a role in the larval development of the roundworm (Lee *et al.*, 1993). This small molecule turned out to be just one entity of a new class of ncRNAs that is present in many, if not all, animals and plants. In the year 2000 the second miRNA, let-7, was found in the same species. This event encouraged researchers to systematically search for more members of this molecule family in *Caenorhabditis elegans*, *Drosophila melongaster* and lastly also in *Homo sapiens* (Lagos-Quintana *et al.*, 2001; Lau *et al.*, 2001; Lagos-Quintana *et al.*, 2001; Lee and Ambros, 2001). The search for new miRNA genes was supported by bioinformatics tools and algorithms which assessed sequence and structural features that may give rise to the characteristic precursor of miRNAs. MiRNAs were found to be highly conserved in mammals and homologs exist in invertebrates also (Carrington, 2003). Thus, many new molecules were identified and experimentally validated. In parallel, their biogenesis and function as post-transcriptional regulators were studied in more detail (Bartel, 2004; Ambros, 2004). Next, tissue specific, as well as developmental stage specific, expression patterns were observed and the data was soon made accessible through public repositories like the miRGator database (Cho *et al.*, 2012; Nam *et al.*, 2008).

With the advancement of high-throughput sequencing technologies, more putative miRNA genes have been detected in human cells. With the 21st release of the miRBase database, the primary repository for miRNA sequences, 1,881 human miRNA hairpin precursor sequences are currently registered and annotated (Kozomara and Griffiths-Jones, 2014; Schmitz and Vearasilp, 2013). From each precursor up to two mature miRNAs can emerge. The miRBase database lists 2,813 human mature miRNAs; a distinctive subset of these is present in all human cell types, and also in human body fluids. Based on the results of different target prediction algorithms miRNAs are estimated to regulate between 30% and 90% of all human protein coding genes (Chang and Mendell, 2007) and consequently many cellular processes. Some predictions suggest an even higher influence on the human genome. The algorithm RNA22 for example predicts miRNA binding sites in the 3′ untranslated region (UTR) of almost every human gene (92.3%) (Miranda *et al.*, 2006). The regulatory efficacy of miRNAs has been demonstrated for cellular processes like differentiation, proliferation, apoptosis and more (Chen, 2004; Bhattacharya *et al.*, 2012; Brennecke *et al.*, 2003; Garzon *et al.*, 2009). In light of this, it seems only natural that deregulation of miRNA expression has been linked to human pathologies.

1.4.1 MiRNA genomic makeup and regulation

Just as in the case of proteins, the genetic code of miRNA molecules resides in the DNA. MiRNA genes are distributed over all human chromosomes and are either intergenic, i.e. they are located in regions between open reading frames of other (protein coding) genes, or they are intragenic, i.e. located in non-translated regions of ORFs called introns. In the first case they have their own promoter region and are independently transcribed, whereas in the latter case they are co-transcribed together with their host genes (Freiesleben *et al.*, 2013). Some miRNAs are organized in clusters, i.e. several consecutive miRNA genes, separated by short nucleotide sequences that reside in an operon on the DNA and thus share single promoter and terminator regions. Deregulated miRNA clusters have been associated with the emergence of solid tumours, such as lung, breast and prostate cancer

(Lai and Vera, 2013). Apart from the transcriptional activation or inhibition through TFs, miRNAs can be the target of epigenetic gene regulation by histone modification or DNA methylation in the same way as other genes (Saito and Jones, 2006).

1.4.2 MiRNA biogenesis

Starting from a gene sequence in the DNA, miRNA biogenesis is the processing system which results in the generation of a maturated short single stranded RNA molecule. This molecule has the potential to bind and repress a mRNA target. In the following, I introduce the canonical miRNA biogenesis pathway as well as alternative forms.

The canonical miRNA biogenesis pathway

In the canonical biogenesis pathway the miRNA gene is transcribed by RNA polymerase II (Pol II) into a long and often polycistronic transcript, referred to as pri-miRNA, which contains a 5′ 7-methylguanosine cap, one or more stem-loop structures and a 3′ poly-(A) tail. Each stem-loop structure with a ~32 nt long imperfectly base-paired stem and a terminal loop (hairpin) can give rise to one or two miRNA molecules. Next, a RNase III enzyme called Drosha and its cofactor DGCR8, together forming the small microprocessor complex, cleave the pri-miRNA molecule into isolated 70-80 nt long stem loop structures referred to as pre-miRNAs. These pre-mature miRNA molecules are recognized by the carrier protein Exportin-5 and are translocated from the nucleus to the cytoplasm. In the cytoplasm, pre-miRNAs are further processed by the RISC loading complex (RLC), composed of the proteins Dicer, TRBP and Ago2. Dicer, another RNase III enzyme, trims the pre-miRNA stem loop into a ~21nt double-stranded miRNA duplex. While the so-called guide strand of this duplex is loaded onto Ago2, the other strand, the passenger, is often degraded. The single stranded mature miRNA, as part of the multi-protein-complex RISC, which emerges from RLC, recognizes and hybridizes with its designated target genes based on imperfect sequence complementarity to binding sites in the 3′ UTR of the target mRNA (Hutvágner and Zamore, 2002; Figure 1.3). The importance of an intact miRNA biogenesis pathway was demonstrated when expression of miRNA processing enzymes Dicer and Ago2 was inhibited and a subsequent up-regulation of miRNA targets could be observed (Rehwinkel *et al.*, 2005; Schmitter *et al.*, 2006).

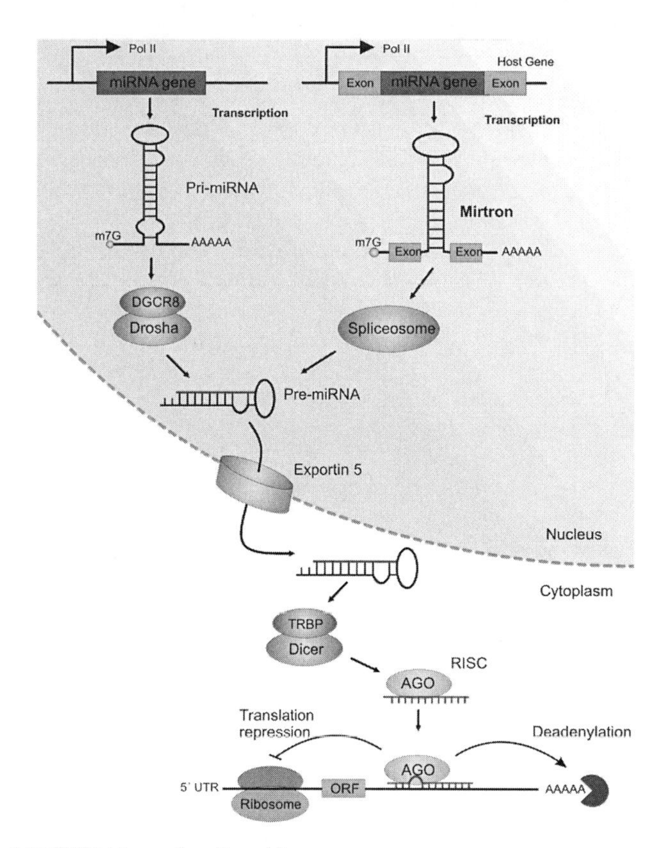

Figure 1.3 MiRNA biogenesis pathway(s).

Endogenous intra- or intergenic miRNA genes are transcribed into so called pri-miRNAs. Pri-miRNAs are cleaved by Drosha (an RNase III enzyme) and its cofactor DGCR8 into single ~80nt long stem-loop structures named pre-miRNA. Pre-miRNAs are transported by Exportin-5 to the cytoplasm, where they are further processed by the Dicer protein complex, which extracts the mature miRNA; a single stranded ~22nt long oligonucleotide. Finally, mature miRNAs are embedded in the RNA-Induced Silencing Complex (RISC) which binds to target mRNA and either induces cleavage (in case of perfect or near prefect sequence complementarity; not shown), suppresses target translation (bottom left) or mediates target deadenylation which results in target destabilisation (bottom right). The figure also illustrates one possible alternative (non-canonical) route for the synthesis of a mature miRNA molecule. In this case the pre-miRNA is a splice product of a miRNA containing host gene. The so-called *mirtron* originates from an intron of the host gene and skips the Drosha-dependent cleavage of pri-miRNA molecules. This kind of miRNA synthesis process is consequently referred to as Drosha-independent miRNA biogenesis pathway. Likewise in the Dicer-independent miRNA biogenesis pathway (not shown) pre-miRNA trimming is skipped.

Alternative miRNA biogenesis pathways

Some splicing products of protein coding genes, referred to as *mirtrons*, resemble the structural features of pre-miRNAs and do not undergo Drosha-mediated cleavage but directly proceed downstream the miRNA processing pathway (Ruby *et al.*, 2007; Figure 1.3). This is one form of a non-canonical miRNA biogenesis pathway which belongs to the class of Drosha-independent pathways. An alternative to that is the Dicer-independent pathway, where the pre-miRNA skips Dicer cleavage and directly proceeds to Argonaute 2 (AGO2) processing by which the guide strand is separated from the stem of the pre-miRNA hairpin loop. The miRNA biogenesis process and its alternative forms are summarized in (Miyoshi *et al.*, 2010) and (Freiesleben *et al.*, 2013).

1.4.3 MiRNA target regulation

Target regulation by single miRNAs

MiRNAs in complex with RISC regulate their targets at the post-transcriptional level by: (A) preventing translation initiation, (B) blocking the translation elongation or causing an early translation termination, or (C) mediating deadenylation of target mRNA (capping of the poly-A tail) which causes destabilisation of the transcript and consequently target mRNA degradation (Figure 1.4).

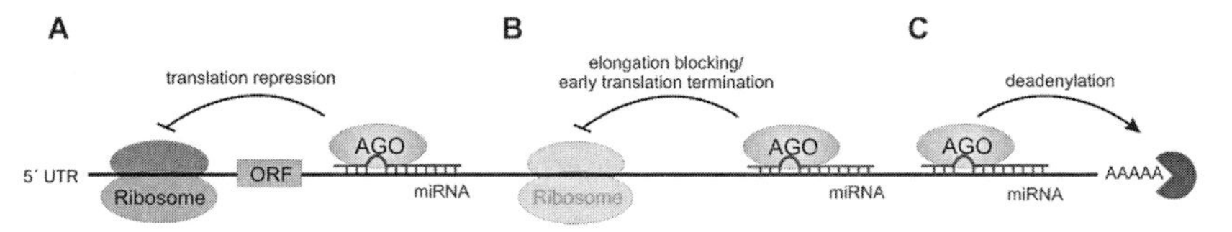

Figure 1.4 MiRNA target regulation mechanisms.
MiRNAs mediate post-transcriptional target gene regulation in different ways, for example by (**A**) blocking translation initiation or (**B**) by inhibiting translation elongation or causing early translation termination. (**C**) Alternatively, miRNAs can mediate accelerated target degradation through mRNA deadenylation and subsequent destabilization. Not shown in this figure is the process of target cleavage which is more common in plants and requires perfect or near perfect sequence complementarity between miRNA and target site.

While target deadenylation and translation repression are the predominant forms of miRNA induced target regulation in mammals, plant genes often contain target sites with perfect or near perfect complementarity to miRNA sequences which facilitates an endonucleolytic mRNA cleavage and therefore an efficient mRNA down-regulation (Yekta, 2004; Jones-Rhoades *et al.*, 2006). All these processes are known under the pseudonym RNA interference (RNAi), and miRNA induced endonucleolytic mRNA cleavage is nowadays used by many experimentalists as a tool for targeted gene silencing (Grimm and Kay, 2007).

Functional target sites in animal gene transcripts, sometimes also referred to as miRNA recognition elements, are typically found in the 3′ UTR of target mRNA. The most critical element in miRNA-target recognition is the seed site to which nucleotides 2-7 of the miRNAs 5′ end often exhibit perfect complementarity (Wang and El Naqa, 2007; Gan and Gunsalus, 2013). Although complementary seed sites can be found in coding sequences, open reading frames and 5′ UTRs, these target sites are usually non-functional (Grimson *et al.*, 2007). However, exceptions are described for example in (Duursma *et al.*, 2008) and (Lytle *et al.*, 2007). Principles of miRNA target regulation have been reviewed in (Engels and Hutvagner, 2006) and (Schmitz, 2013). As mentioned before, functional target sites have, in most cases, only imperfect sequence complementarity with the miRNAs that hybridize to them. The resulting hybrid often includes mismatches, gaps, G:U wobbles and internal loops. This makes miRNA target prediction a challenging task. Many different approaches and algorithms to tackle this problem exist and are described in more detail in the next chapter.

Target regulation by multiple miRNAs and miRNA multiplicity

Growing evidence and many predictions suggest that a mRNA target can be regulated by more than just one miRNA, rather by several collectively and sometimes cooperatively acting miRNAs (Lai *et al.*, 2012; Schmitz *et al.*, 2014; Figure 1.5A). Collective regulation induces stronger target repression. One miRNA, on the other hand, can have binding sites in several, sometimes dozens or more, target genes. This phenomenon is referred to as target multiplicity and is caused by the fact that perfect sequence complementarity is only required in the seed region (6-8nt in length) in a functional miRNA/mRNA hybrid (Figure 1.5B). Furthermore, mismatches in this region can be compensated by extensive base pairing at the 3′ end of the miRNA. These target sites are also referred to as 3′ compensatory sites (Lim *et al.*, 2005; Friedman *et al.*, 2009). Target multiplicity has the consequence that the regulatory capacity of an overexpressed miRNA is distributed among several targets and thus often induces only mild repression on a single mRNA. Interestingly, one-to-many and many-to-one target regulation mechanisms have been described for transcriptional target regulation by TFs before (Shalgi *et al.*, 2007).

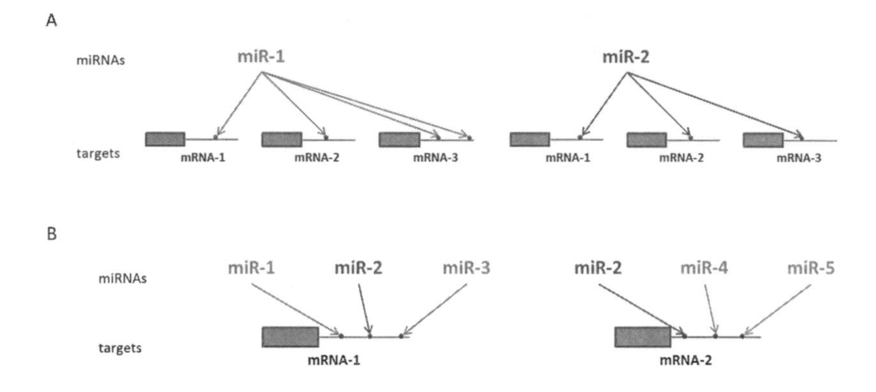

Figure 1.5 Multiple and collective target regulation.
(**A**) MiRNAs can have binding sites in multiple targets and therefore may induce only mild repression in individual targets. (**B**) Targets are often regulated by several miRNAs. Those genes being targeted by 15 or more different miRNAs are also referred to as miRNA target hubs.

Indirect target regulation and regulatory networks

In indirect target regulation, miRNAs cause expression changes of non-miRNA target genes (Hill *et al.*, 2014). This can often be observed in high-throughput transcriptomics experiments upon miRNA transfection or silencing (a.k.a. miRNA perturbation experiments). Indirect gene regulation by miRNAs may be caused through miRNA-repressed activators or inhibitors of indirect targets, e.g. TFs (Selbach *et al.*, 2008; Tu *et al.*, 2009). Therefore, indirect miRNA mediated target regulation involves at least three components: (i) a miRNA, (ii) an intermediary miRNA target, and (iii) an indirect target gene. These components are often found to be arranged in small regulatory circuits or larger regulatory networks. The analysis of such networks is a current topic of active research. The computational analysis of such regulatory circuits is discussed in detail in Chapter 3.

MiRNA cooperativity

Recently, another feature of collaborative target regulation by miRNAs was uncovered. It was demonstrated that under specific circumstances, pairs of miRNAs can cooperatively regulate mutual targets genes (Vella *et al.*, 2004; Hon and Zhang, 2007). An optimal distance of 13-35nt between seed sites of cooperating miRNAs is required to induce a synergistic effect and thereby a more efficient target repression (Saetrom *et al.*, 2007; Figure 1.6). This sophisticated mechanism, beyond the enhanced repressive effect, can facilitate a fine-tuned target regulation and can act as an efficient buffer to transcriptional noise (Lai *et al.*, 2012). Furthermore, the cellular demand for an adjustment of gene expression, e.g. in different cellular programs, can be supported by cooperative target regulation (Lai *et al.*, 2012, 2013; Schmitz *et al.*, 2014). Gene expression dynamics induced by collective and cooperative target regulation are analysed in Chapter 3, while a workflow for the prediction of cooperating miRNA and their mutual targets genes is described in Chapter 4.

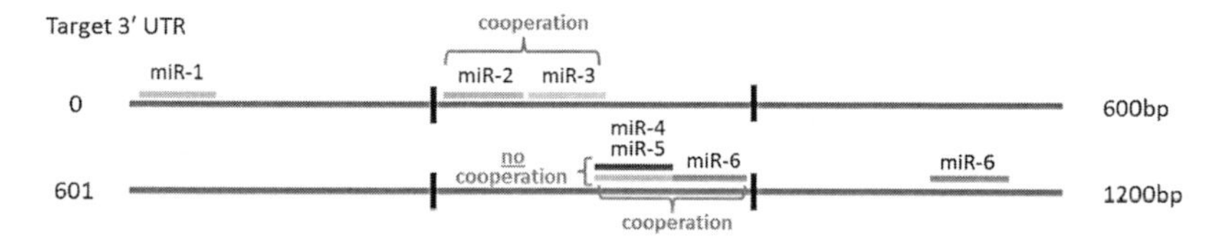

Figure 1.6 MiRNA pairs may cooperate when their target sites are in close proximity.
In Saetrom *et al.* (2007) an optimal distance (13-35nt) of target seed sites was found to facilitate a cooperative effect on target repression by pairs of miRNAs.

Target repression efficiency

The degree to which a miRNA can control the expression of its target can vary significantly. While, for example, Selbach and colleagues (Selbach *et al.*, 2008) have shown that miRNAs induce mild expression changes in a large number of target genes, other studies demonstrate strong target repression upon direct miRNA transfection or in reporter gene experiments (Gabriely *et al.*, 2008; Jiang *et al.*, 2009a). In fact, the efficiency of target repression by miRNA regulation is determined by a number of factors which are not yet understood in detail (Filipowicz *et al.*, 2008). However, target repression efficiency can be determined experimentally through transcriptomics approaches, e.g. by RT-PCR[3], Microarray or RNAseq[4], and proteomics approaches, e.g. immunoblotting[5] or SILAC[6]. Investigators often oversee the fact that transcriptomics approaches can capture only those cases in which miRNA induces a reduction in target mRNA concentration, either by mediating target deadenylation or directly through target cleavage. However, the effect of those cases in which a miRNA represses translation (Figure 1.4) can only be measured on the post-translational level by proteomics approaches. All these methods, however, carry the risk of measuring unintended effects induced by other uncontrolled factors, e.g. regulatory loops, off-targets, etc. Hence, the most adopted approach for identifying the effects of miRNAs on target gene expression is the reporter gene assay. For example, Mukherj and colleagues adopted a two-color

[3] RT-PCR - **R**everse **T**ranscription **P**olymerase **C**hain **R**eaction; a technique for measuring RNA expression via the amplification of the reverse transcript (cDNA)

[4] RNAseq - RNA sequencing; a specific type of next generation sequencing (a.k.a. deep sequencing)

[5] Immunoblotting – a.k.a. Western blot; a technique based on gel electrophoresis for quantifying protein expression in a sample

[6] SILAC - **S**table **I**sotope **L**abelling with **A**mino acids in cell **C**ulture; a mass spectrometry based technique for quantitative protemics (Selbach *et al.*, 2008a)

fluorescent reporter system in a single-cell analysis and found that, although the average miRNA-induced protein expression is modest, it underlies an extreme variation among different individual cells (Mukherji *et al.*, 2011). Furthermore, they found that below a certain threshold level of mRNA the target protein synthesis is highly repressed by miRNAs and that it responds sensitive to mRNA concentration changes near the threshold. These results are reflected in their mathematical model which also showed that with increasing abundance of target mRNA, the availability of miRNAs for repression decreases. The strength of the interaction between miRNA and its target and their relative abundance decides the sharpness of the switch from full repression to escape from miRNA repression (Mukherji *et al.*, 2011).

1.4.4 MiRNAs in human pathologies

MiRNAs have been associated with all kinds of human diseases, ranging from autoimmune diseases, cardio-vascular diseases, neurodegenerative diseases, metabolic diseases to cancer (Thum *et al.*, 2008; Farazi *et al.*, 2013). The miR2Disease database, a manually curated database that is a resource for miRNA deregulation in human pathologies, contains in its current release 349 miRNAs associated with 163 diseases (release April 2008; Jiang *et al.*, 2009). Another resource for miRNA-disease associations contains even more entries (HMDD v2.0 database; 572 miRNAs; 378 diseases; Li *et al.*, 2013).

In our own efforts, we demonstrated:

- the important role of miRNAs in the regulation of cancer signalling pathways (Nikolov *et al.*, 2010),
- that the miRNA miR-195 regulates the cell cycle checkpoint kinase WEE1 in malignant melanoma (Bhattacharya *et al.*, 2012),
- that DNp73 suppression of miR-205 expression can mediate anti-cancer drug resistance and facilitate E2F1 accumulation which leads to apoptosis resistance and the emergence of aggressive tumour cells (Alla *et al.*, 2012; Vera *et al.*, 2013),
- the oncogenic potential of miR-638, which is capable of protecting melanoma cells from apoptosis and autophagy, and thereby promoting melanoma metastasis (Bhattacharya *et al.*, 2015), and
- that the miR-224/miR-452 cluster targets the metastasis suppressor TXNIP and thereby mediates E2F1-induced EMT and invasion in malignant melanoma (Knoll *et al.*, 2014).

Another important point in the context of miRNA-disease associations is the fact that miRNAs are a valuable source for biomarkers used for the diagnosis and prognosis of many human pathologies (Yanaihara *et al.*, 2006; Abeel *et al.*, 2009; Ferracin *et al.*, 2010; Zhao *et al.*, 2010; Bonfrate *et al.*, 2013; Barbarotto *et al.*, 2008). In many diseases, miRNAs are found to be differentially expressed compared to healthy controls. Sometimes one can even observe distinct miRNA expression profiles through the course of disease progression (Ueda *et al.*, 2010; Bhattacharya *et al.*, 2015). The most stable and reliable miRNA biomarkers have been found in human body fluids, especially blood, plasma and serum, in the form of circulating miRNAs (Chen *et al.*, 2008; Etheridge *et al.*, 2011; Kosaka *et al.*, 2010; Zampetaki *et al.*, 2011; Mitchell *et al.*, 2008). Additionally, miRNAs became a popular subject for the design of novel therapeutic interventions. For details see the reviews by Seto (2010), and Kasinski & Slack (2011).

In the next chapter bioinformatics and systems biology approaches used for studing the role and functioning of miRNAs are introduced and critically discussed. We integrated and enhanced some of these approaches for our analyses which are described in detail in the subsequent chapters.

2 A comparative analysis of methods in computational miRNA biology

This chapter is based on the following publications:

Schmitz U, Vearasilp K (2013) miRBase. In Encyclopedia of Systems Biology (ed. Dubitzky W, Wolkenhauer O, Cho K, Yokota H), pp. 1363-1366. Springer New York.

Schmitz U, Wolkenhauer O (2013) Web resources for microRNA research. Adv Exp Med Biol, 774, 225–250.

Vera J, Lai X, **Schmitz U**, Wolkenhauer O (2013) MicroRNA-Regulated Networks: The Perfect Storm for Classical Molecular Biology, the Ideal Scenario for Systems Biology. Adv Exp Med Biol, 774, 55–76.

Vera J, **Schmitz U** (2013) Computational MicroRNA Biology. In Encyclopedia of Systems Biology (ed. Dubitzky W, Wolkenhauer O, Cho K, Yokota H), pp. 473-480. Springer New York.

Vera J, Wolkenhauer O, **Schmitz U** (2014) Cancer Systems Biology: Current Achievements in – Omics Data Analysis, Network Reconstruction and Mathematical Modelling. In eLS. John Wiley & Sons Ltd, Chichester.

Synopsis

In this chapter I compare and discuss computational approaches and methodologies used in miRNA biology research with a focus on those that were adopted in this thesis. More specifically, I critically discuss methods for the prediction of miRNAs and their targets genes; I introduce miRNA structural prediction and miRNA expression profiling approaches coupled with the functional characterization of miRNA sets (e.g. differentially expressed miRNAs). Thereafter, I discuss available web resources developed to share data for miRNA research. Furthermore, I review modelling approaches used for miRNA biology and describe models of individual miRNA-target interactions, small regulatory circuits, and large regulatory networks. Finally, I introduce case studies where systems biology approaches have been employed for studying miRNA regulation in human diseases. While describing bioinformatics and systems biology approaches separately in this chapter, I explain in the subsequent chapters how we integrated these two complementary research fields to achieve our goals.

2.1 Bioinformatics approaches

In this section I give an overview of approaches, methodologies, tools and resources designed for and used in the field of computational miRNA biology. While there are many possible applications of bioinformatics methods for investigating miRNA related questions, I focus here on the four most widely used applications: (i) miRNA gene prediction, (ii) miRNA target prediction, (iii) miRNA expression profiling, and (iv) web-based tools and databases for miRNA research.

2.1.1 Computational identification of miRNA genes

MiRNAs can be identified experimentally using, for example, northern blotting or deep sequencing. The latter has the highest sensitivity among the detection methods and recently provided a giant leap in the number of identified miRNAs (Kozomara and Griffiths-Jones, 2011). The computational prediction of miRNA genes, on the other hand, relies on methods designed to identify loci in the genome from which mature miRNA sequences may arise. The predictions of these methods are obtained from the analysis of several sources of data, including multiple genome comparisons and transcriptomics data. While the first miRNA genes were predicted solely by computational algorithms, nowadays these predictions are used to identify miRNAs in uncharacterised reads of deep sequencing experiments (Friedländer *et al.*, 2008). Furthermore, they are used to detect miRNAs with low expression levels which are hard to identify with purely experimental approaches or those expressed in few tissues only (Herbig and Nieselt, 2013).

The development of methods for the *in silico* detection of new miRNAs benefitted from efforts aiming to understand the genetic context and biosynthesis of miRNAs. It is currently known that miRNA genes are often located in intergenic regions with their own promoters and terminators or are organized in clusters to share *cis*-regulatory elements[7] (Freiesleben *et al.*, 2013). Alternatively, many miRNAs originate from gene introns and their expression is co-regulated with that of their host gene. Consequently, miRNA prediction algorithms start by identifying conserved sequences in these genomic regions by applying multiple sequence alignment algorithms such as ClustalW (Larkin *et al.*, 2007).

MiRNA precursors are 70-80 nt (nucleotides) long and form characteristic stem loop structures (Figure 2.1). Therefore, the second step in many miRNA gene prediction algorithms is to predict the secondary structure that may arise from conserved intergenic or intronic subsequences. This is done with the help of RNA secondary structure prediction algorithms, such as mfold (Zuker, 2003) or RNAfold, which is part of the Vienna package for RNA structure prediction (Lorenz *et al.*, 2011). These algorithms

---

[7] A *cis*-regulatory element is a DNA sequence that is involved in the transcriptional regulation of a gene. These sequences are often transcription factor binding sites that are located in a gene promoter region.

determine the minimum free energy (MFE) structure from RNA sequences based on possible base pairing and base stacking interactions. For this purpose, most RNA structure prediction tools rely on a dynamic programming algorithm. Candidates are then filtered for those that display typical pre-miRNA-like stem loop structures with few bulges or inner loops and for conserved structures (Mendes *et al.*, 2009). Some algorithms filter a candidate set by keeping only the thermodynamically most stable and structurally robust precursors. Structure stability is measured in terms of MFE of the fold and structural robustness is given in case the stem loop structure is preserved despite small mutations in the sequence (Mendes, 2013). The identification of miRNA pre-cursors characteristic structures have been addressed by some tools which use machine learning approaches such as support vector machines or hidden Markov models for the classification of RNA secondary structures. These tools use known pre-cursor structures as positive training set and randomly generated stem loop structures as negative training set (Mendes, 2013).

Figure 2.1 Characteristic hairpin loop structure of a miRNA precursor.

Newly sequenced genomes often benefit from known miRNAs in other species. The identification of homologue sequences provides an initial set of miRNA candidates for these genomes. In the last decade a number of prediction algorithms have been published which has led to the identification of hundreds of new miRNAs. Some of these algorithms are listed in Table 2.1.

Method	Algorithm	Reference (PMID)
Comparative methods	MiRScan	12672692
	DIANA-microH	19765283
	RNAmicro	16873472
	MiRPred	18088431
	MiRFinder	17868480
	MiRAlign	15994192
	MirSeeker	23153516
Machine learning-based methods	triplet-SVM	16381612
	MiPred	17553836
	ProMir	16845048
miRNA prediction from RNAseq data	miRDeep	18392026
	miRTRAB	20370911

Table 2.1 MiRNA (pre-cursor) prediction algorithms.
We developed our own ncRNA gene prediction software, named *moses* (Raasch *et al.*, 2010), which integrates different approaches but is not specific to miRNAs, which is why it is not listed in this table.

2.1.2 Approaches for the prediction and analysis of miRNA target genes

MiRNA target prediction

The computational prediction of miRNA targets in plants is simpler than in animals. In plants, most functional miRNA target interactions are based on near perfect or perfect sequence complementarity of a miRNA to a site in the ORF (the coding region) of a gene. In these cases the targets undergo miRNA-mediated cleavage (Rhoades *et al.*, 2002; Schwab *et al.*, 2005). Potential target genes can be easily detected by identifying sites in the ORFs that match the reverse complement of plant miRNAs. Most plant miRNA target prediction tools utilize BLAST (Basic Local Sequence Alignment Search Tool; Altschul *et al.*, 1990) for this purpose. Animal miRNAs, in contrast, preferentially bind to the 3′ UTR of mRNAs but target sites may also occur in the ORF or the 5′ UTR. Moreover, in most cases only imperfect sequence complementarity is required for functional miRNA-target interactions, i.e. often complementary hexamers or heptamers are sufficient for target recognition. Considering this, each miRNA is in principle able to hybridize with thousands of target genes and may consequently regulate their expression. For example, given that the average 3′ UTR length of mammalian genes is ~1,000 nt (Hendrickson *et al.*, 2009) a complement for a given heptamer would presumably be found in every 4^7 (16,384) bases, i.e. in at least every 17th gene (Schmitz, 2013). This would mean, in *homo sapiens* (~19,000 genes) each miRNA could regulate more than 1,100 genes. Indeed, it was shown that miRNAs induce '*widespread changes in protein synthesis*' (Selbach *et al.*, 2008a), but in reality the number of targets per miRNA ranges between only a handful and several hundred (Hafner *et al.*, 2010). Therefore, the main challenge for prediction algorithms is to reliably identify functional miRNA-target pairs. Moreover, it is desired to filter the most efficiently regulated targets for a given miRNA or the most efficient regulator for a given target.

The first target prediction algorithms have been developed based on the few functional miRNA-target interactions known at that time (Enright *et al.*, 2003). Then as now, most of the methods were based on: (a) sequence complementarity between a miRNA and its target (with focus on the seed region), (b) thermodynamic stability of the miRNA-mRNA duplex (characterized by the hybridization energy), (c) evolutionary conservation of target sites, (d) compositional and sequence features of the target site, and (e) homologous target sites in related species.

As of today, a plethora of miRNA target prediction algorithms which use different computational means for their predictions have been proposed (Ritchie *et al.*, 2013). Many of these use secondary structure prediction based on a dynamic programming approach to determine possible miRNA-mRNA duplex structures and predict duplex hybridization energies. The underlying assumption is that stable duplexes are more likely to be functional. Some algorithms employ machine learning-based classifiers trained with sequence and structural patterns of validated miRNA-target pairs and/or target expression profiles from miRNA transfection/knockout experiments. Furthermore, based on selected features, miRNA target prediction algorithms apply different scoring schemes that are used to mirror the likelihood or confidence on a predicted miRNA target pair.

In total, more than three dozen algorithms exist that can predict potential miRNA binding sites in mRNA targets (Table 2.2). However, many false positive predictions are produced by these algorithms (Chi *et al.*, 2012). The reason is that many of them are biased due to a limited and homogenous set of validated miRNA target interactions used to design or train an algorithm. Furthermore, there is no clear best performing algorithm because also the test sets used in benchmarks are often homogeneous and limited to those miRNA target interactions that have been validated based on predictions from the early developed algorithms. Therefore, it is no surprise that e.g. predictions from miRanda (Enright *et al.*, 2003), one of the first algorithms proposed, largely overlap with those of more recent target prediction algorithms. In fact, miRanda predictions have the largest relative overlap with predictions from other algorithms (Ritchie *et al.*, 2009). Two reviews on miRNA target prediction algorithms are provided by Sethupathy *et al.* (2006a) and (Ritchie *et al.*, 2013). See an overview of the existing algorithms sorted by the year in which they were developed in Table 2.2.

2003	2004	2005	2006	2007	2008	2009	2010	2011	2012	2013	2014
TargetScan	RNAhybrid	TargetBoost	RNA22	NBmiRTar	MirTarget2	DIANA-microT 3.0	T-Rex	RepTar	miRmap	miMsg	TargetS
miRanda	DIANA-microT	TargetScanS	MicroTar	GenMir++/3	FindTar	TargetMiner	TargetSpy	mirSOM		microMUMMIE	
		PicTar	MiTarget	PITA	mirWIP		SVMicrO			PACCMIT-CDS	
		MovingTargets	MirTarget	TargetRank	MirTif		mirSVR				
		MicroInspector		ElMMo	Sylamer						
		GenMir		StarMir							
				mRTP							

Table 2.2 Timeline of miRNA-target prediction algorithms.
This overview is an expansion of our previously published summary in (Schmitz and Wolkenhauer, 2013).

Experimental target validation

Reliable identification of miRNA-target pairs requires experimental support in any event. One way of doing this is by transfecting cells with mature miRNAs or adenovirus vectors that induce overexpression of a certain miRNA. Subsequent mRNA expression experiments (e.g. microarray, RT-PCR or RNAseq) may give rise to miRNA induced target regulation, especially in case a predicted target gene is down-regulated. This approach, among others, led to the validation of many miRNA-target interactions which can now be found in databases of experimentally validated targets, e.g. TarBase and miRTarBase (Vergoulis *et al.*, 2011; Hsu *et al.*, 2013).

Another widely used approach to support computationally predicted targets for specific cells or tissues is done by performing transcriptomics experiments for both miRNAs and mRNAs. Inversely correlated miRNA-target pairs may reflect miRNA induced target repression (Lim *et al.*, 2005; Farh *et al.*, 2005; Huang *et al.*, 2007). Expression correlation is considered as feature in some of the recently developed data driven target classification algorithms, as for example GenMir3, MirTarget2, and T-REX (Huang *et al.*, 2007; Wang and El Naqa, 2007; Volinia *et al.*, 2009). However, transcriptomics-based evidence does not account for any direct interaction nor does it capture cases of miRNA-mediated translation repression (Bhattacharya and Kunz, 2013). MiRNAs can repress translation without detectable differences in target mRNA levels (Filipowicz *et al.*, 2008). Therefore, immunoblotting experiments are often employed to confirm miRNA-induced target repression at the post-translational level.

Selbach and colleagues recently developed a high-throughput proteomics approach called pSILAC (pulsed stable isotope labelling with amino acids in cell culture) that directly measures genome wide changes in protein synthesis in response to changes in miRNA expression (Selbach *et al.*, 2008). However, proteomics approaches cannot differentiate between direct and indirect target regulation. The expression of genes is sometimes altered by transcriptional effectors, e.g. TFs, that are repressed by a miRNA (Bhattacharya and Kunz, 2013). Therefore, target specificity is often determined through reporter assays that express a putative target gene either with or without intact 3′ UTR. The underlying assumption is that the miRNA cannot suppress the target in case of a lacking 3′ UTR (Bhattacharya and Kunz, 2013). Although this technique provides evidence for the miRNA-target gene interaction and narrows it down to the 3'UTR region, it fails to illustrate the specific target site. This problem can be addressed by site-directed mutagenesis experiments in which miRNA binding sequences are mutated and analysed in reporter gene assays (Bhattacharya and Kunz, 2013). A mutated target site prevents a miRNA from hybridizing with its target and therefore prevents target repression.

Another recently developed approach provides strong evidence for RISC hybridization with a target mRNA. The HITS-CLIP (high-throughput sequencing of RNAs isolated by crosslinking immunoprecipitation) method developed by Licatalosi *et al.* (2008) can be used to identify Argonaute-mRNA as well as Argonaute-miRNA hybrids through crosslinking immunoprecitpitation. The sequence of the binding site and the identity of the miRNA are determined through RNAseq. This method allows the exact localization of miRNA binding sites in the target sequence. Alternatives to this approach are the Photo Activatable Ribonucleoside enhanced CLIP (PAR-CLIP) which has been proposed by Hafner *et al.* (2010) and Crosslinking, Ligation, and Sequencing of Hybrids (CLASH) proposed by Helwak *et al.* (2013).

Prediction of target repression efficiency

Several approaches for the prediction of the target repression efficiency, i.e. the reduction of target protein concentration upon miRNA regulation, have been proposed. However, none have tested the performance in an unbiased way. Certain features of a predicted miRNA-target hybrid are considered to impact upon the target repression efficiency, e.g. the target site accessibility, the target site location, the number of other sites in the same target and in other targets, and the thermodynamic stability (binding energy) of the duplex. As indicated before, one cannot say with certainty to which degree any particular feature influences target repression efficiency (Schmitz, 2013).

In (Lai *et al.*, 2013) we combined miRNA target prediction with mathematical modelling to simulate target repression dynamics as a consequence of post-transcriptional regulation by single or multiple miRNAs. The design and application of the hybrid approach is described in detail in Chapter 3.

2.1.3 Structure prediction and analysis of single and interacting RNA strands

An approach which is not specific to computational miRNA biology, but is often applied in this context, is the prediction of RNA secondary structures and their analysis. For example, RNA secondary structure prediction is used in the computational identification of novel non-coding RNAs (including miRNAs; Meyer, 2007). It is assumed that conserved and thermodynamically stable RNA structures are likely to have a function in the cell (Willingham, 2005). In the case of miRNA gene identification a widely used approach is to look for conserved subsequences in inter- and intragenic regions, or for homologues and paralogues of known miRNA genes. Alternatively, putative and known transcripts are scanned for conserved secondary structures and occurrences of characteristic ~80nt long hairpins (Figure 2.2) that may form pre-cursors of previously unknown miRNAs after Drosha processing of the primary transcript (pri-miRNA; Mendes, 2013). Many miRNA precursors have accordingly been predicted and only later experimentally validated. In our own software framework, named *moses*, we have integrated RNA secondary structure prediction along with other methods (e.g. sequence alignment, structural homology search, etc.) for an accurate and reproducible prediction of non-coding RNA genes (Raasch *et al.*, 2010).

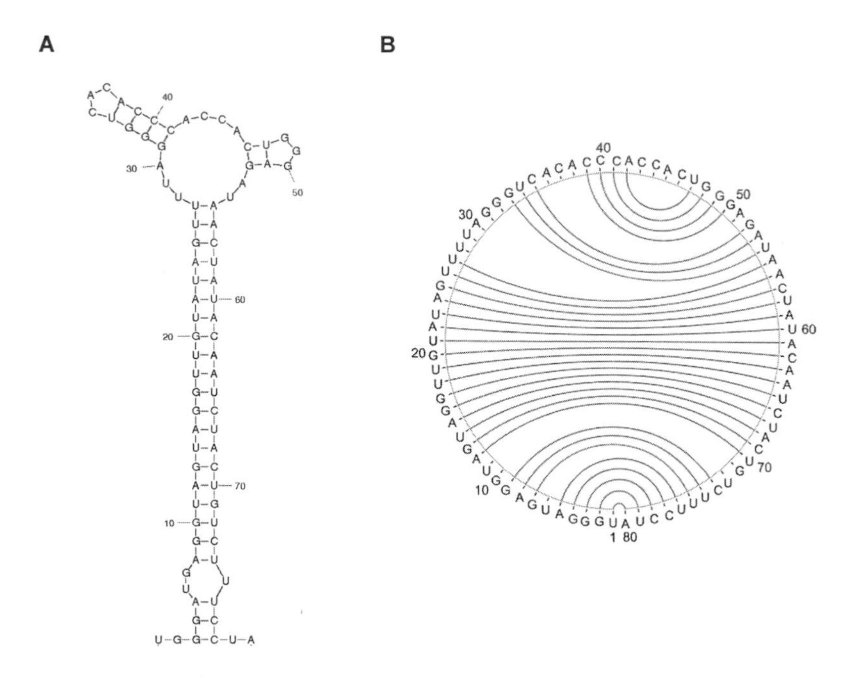

Figure 2.2 Predicted secondary structure for the miRNA precursor let-7a-1.
A: Linear representation (*ΔG = -33.48 kcal/mol*); **B:** Circular structure plot. The secondary structure was predicted and visualized using the mfold web server (Zuker, 2003).

Moreover, prediction of the secondary structure of RNA duplexes has been employed in miRNA target prediction ever since the earliest algorithms (Enright *et al.*, 2003). One approach for finding the optimal folding of a single RNA strand or a duplex composed of two RNA sequences is to computationally identify the structure with maximum base pairs. This poses an optimization problem which can be most efficiently solved by using a dynamic programming approach (Eddy, 2004). In the simplest form of such an approach, a score 1 is awarded for base pairs and 0 otherwise. Considering, for example, a single RNA sequence, the algorithm systematically determines optimal folding for sub-sequences $[i..j]$, that is, the maximum number of nested base pairings this sub-sequence can establish. There are four ways that this structure can be constructed (Figure 2.3A):

A i,j form a base pair, that adds to the structure $i+1..j-1$
B i is unpaired and adds to the structure $i+1..j$
C j is unpaired and adds to the structure $i..j-1$
D i and j form pairs with other bases, thereby the structure for $i..j$ is the joint of two sub-structures established by the subsequences $i..k$ and $k+1..j$. This construct is referred to as *bifurcation*.

The optimal score $S(i,j)$ for each (sub-)sequence is a function of optimal scores of smaller sub-sequences (Eddy, 2004). Based on the four ways the structure of a (sub-)sequence is established, $S(i,j)$ is determined as follows:

$$S(i,j) = max \begin{cases} S(i+1,j-1)+1 \quad [if\ i,j\ \ base\ pair] \\ S(i+1,j) \\ S(i,j-1) \\ max_{i<k<j} S(i,k) + S(k+1,j) \end{cases}$$

The score is recursively computed for each sub-sequence which is illustrated in the triangular matrix in Figure 2.3B, which is initialized by assigning score 0 for sub-sequences of length 0 and 1 (Figure 2.3B, left). The matrix is then systematically filled with scores for the step-wise growing sub-sequences. The optimal score for the complete sequence appears in the upper right corner (Figure 2.3B, middle). One can now trace back the route that led to this optimal score and thereby determine the optimal fold (Figure 2.3B, right).

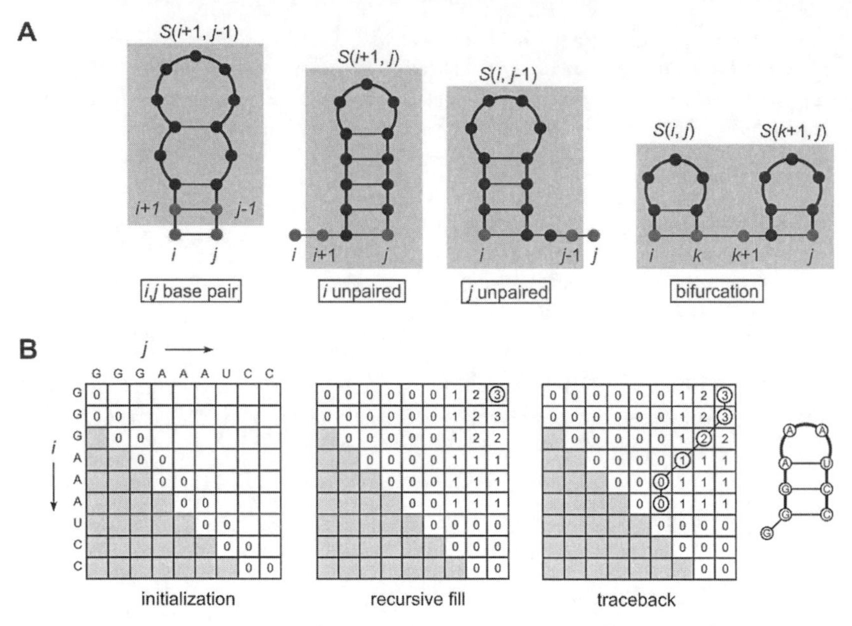

Figure 2.3 Dynamic programming algorithm used for RNA secondary structure prediction. **A:** Four ways the folding for sequence $[i..j]$ can be established. The optimal score $S(i,j)$ is recursively found by identifying the one with the most base pairs among these four possibilities. Grey dots indicate bases that are added to the sub-sequence. **B:** Illustration of the dynamic programming algorithm showing the scoring matrix for the RNA sequence GGGAAAUCC. The matrix is initialized by assigning score 0 for sub-sequences of length 0 and 1 (left) and thereafter recursively filled with scores for the optimal sub-sequences of length > 1 (middle). Finally, by tracing back the path that led to the maximum score the optimal secondary structure can be determined (right). Illustration inspired by (Eddy, 2004).

The computation time for determining the structure with maximum base pairs is proportional to N^3 (N being the length of the sequence).

Today's secondary structure prediction algorithms, however, are based on the assumption that every system (including RNA complexes) tries to achieve a minimum energy. This in turn means that a reaction or fold that releases energy, e.g. through duplex formation, is favoured. A quantitative measure for the release or consumption of energy in a reaction is the change ΔG of the Gibbs energy. A negative Gibbs energy change ($\Delta G < 0$) suggests an exergonic reaction, i.e. one that releases energy and thus will be favoured. The overall free energy of a RNA structure is determined by the sum of terms for different loops, base pairing interactions and base stacking (Eddy, 2004). However, so called energy minimization algorithms still use a similar dynamic programming approach, as described above, to find minimum free energy structures.

Structural element	ΔG	Information
External loop	0.00	0 ss bases & 1 closing helices
Stack	-3.30	External closing pair is G_1-C_9
Stack	-1.50	External closing pair is G_2-C_8
Helix	-4.80	3 base pairs
Hairpin loop	4.00	Closing pair is G_3-U_7

$\Delta G = -0.80$

Table 2.3 Details on thermodynamic terms for a minimum free energy structure.
Results for the predicted secondary structure of sequence GGGAAAUCC received from the mfold web server which is based on an energy minimization algorithm (Zuker, 2003). On the right side is an illustration of the predicted secondary structure which differs from that predicted by a base pair maximization algorithm (Figure 2.3).

Thermodynamically stable structures can be predicted with tools like mfold, RNAz and RNAfold (Zuker, 2003; Washietl *et al.*, 2005; Hofacker, 2003) which compute the minimum free energy (MFE) structures of a single RNA sequence or a putative RNA duplex composed of a miRNA and its designated target mRNA. A miRNA-mRNA duplex also requires an energetically favourable and stable structure, i.e. a strongly negative ΔG. In the first miRNA target prediction algorithm proposed, Stark *et al.* (2003) applied structure prediction by energy minimization to predict miRNA targets in *D. melanogaster* (fruit fly). They used the mfold software, which was originally designed for structure prediction of single nucleic acid sequences, and therefore they concatenated the potential target-site and the miRNA sequence to a continuous sequence string by using a short standard hairpin-forming linker sequence 'GCGGGGACGC' (Figure 2.4). The same strategy was also used in Enright *et al.* (2003), Rajewsky and Socci (2004), and Lewis *et al.* (2003).

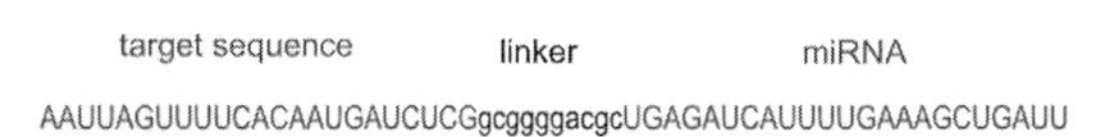

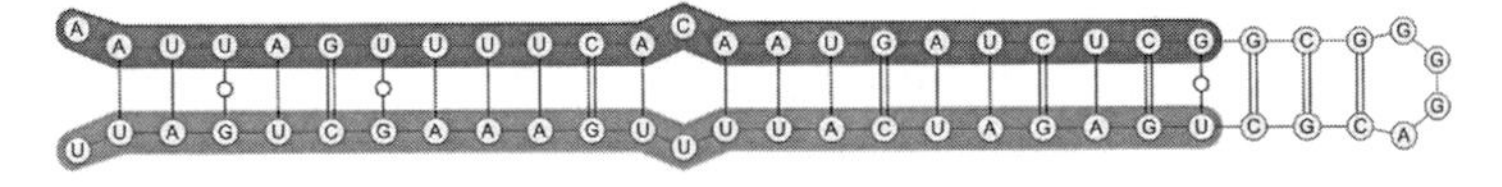

Free energy of secondary structure: -34.20 kcal/mol

Figure 2.4 Secondary structure of concatenated target site and miRNA sequences.
A hairpin-forming linker sequence was used to concatenate the two sequences into a single.

In the new generation of RNA structure prediction tools, some tools were especially designed for RNA hybrids, e.g. RNAcofold and RNAduplex (Hofacker, 2009; Lorenz *et al.*, 2011). The former allows intra- and inter-molecular base paring, while the latter allows only inter-molecular base paring for energy calculation. Principles of RNA duplex structure prediction were adopted and customized for the case of miRNA-target mRNA interactions in a tool called RNAhybrid which was optimized for finding small RNA molecules that bind to large RNA molecules. The program searches for the energetically most favourable hybridization site of miRNA in mRNA, thereby not allowing intra-molecular hybridizations. Some algorithms also predict the secondary structure of the full length 3′ UTR (Robins *et al.*, 2005; Hammell *et al.*, 2008) or even the entire mRNA (Thadani and Tammi, 2006) for the estimation of site accessibility and calculation of the energetically favourable arrangement (with or without miRNA). However, predicted structural folding of 3′ UTR sequences in isolation may differ from their folding in whole mRNA secondary structure predictions due to possible base pairs forming between 3′ UTR and coding region or 5′ UTR. Moreover, it has been shown that the prediction accuracy decreases with sequence length (Wang and Wang, 2006).

Other tools in contrast compare the free energy of the potential hybrid with that of possible structure formations of the free mRNA 3′ UTR. Intra-molecular hybridization of mRNA might make potential miRNA binding sites hard to access and thus target regulation less efficient. Therefore these tools, e.g. TargetMiner (Bandyopadhyay and Mitra, 2009) and PITA (Kertesz *et al.*, 2007), compute $\Delta\Delta G$, which is the difference between the free energy gained by establishing a miRNA/mRNA duplex (ΔG_{duplex}) and the free energy consumed by unpairing the target site nucleotides (ΔG_{open}):

$$\Delta\Delta G = \Delta G_{duplex} - \Delta G_{open}$$

2.1.4 MiRNA expression analysis and functional characterization

MiRNAs regulate gene expression and thereby alter gene functioning. Individual or concerted miRNA action therefore has consequences for biological processes, which are thereby triggered, altered or suppressed. In an extreme case, deregulated miRNAs can mediate or even cause diseases. The importance of investigations into miRNA expression and function can therefore not be understated.

MiRNA expression profiling

The expression of miRNAs can be, similar to mRNAs, determined in different ways; either individually through RT-PCR or in a high-throughput manner through HT-qPCR (high-throughput quantitative PCR), microarrays or RNAseq. The raw data generated in these high-throughput experiments have to undergo pre-processing steps before statistical methods can be applied for the identification of differentially expressed miRNAs. These involve quality control, filtering and normalization.

As part of the quality control in microarray experiments artefacts (e.g. printing or washing defects) are identified on the images by visual inspection or comparison to arrays from the same sample set. In case of HT-PCR data, software tools check if probes were evenly amplified after a pre-defined number of cycles. The normalization step is subject to debate in the scientific community. Different approaches exist for normalizing microarray data (e.g. Robust Multi-array Average (RMA), Affymetrix MAS5, Factor Analysis for Robust Microarray Summarization (FARMS), etc.) and for HT-PCR data (e.g. quantile normalization, normalization based on rank invariant features, or normalization based on reference or housekeeping genes (a.k.a. $\Delta\Delta Ct$ normalization)). In a large-scale assessment of miRNA HT-PCR normalization methods, the geometric mean normalization turned out to be the most stable and reliable method (Mestdagh *et al.*, 2009). In this method the mean expression value of each sample is derived based on the number of PCR cycles (cycle threshold = Ct) per feature. Thereafter all Ct values are scaled based on the relation of the sample mean to the mean of the other samples. In the case of RNAseq, fragmented sequence reads have to be assembled. This requires sophisticated bioinformatics approaches that involve steps of (i) pre-processing (filtering), (ii) graph construction & simplification, and (iii) post-processing (El-Metwally *et al.*, 2013).

After pre-processing, statistical analyses including multiple testing correction and clustering are typically applied in miRNA profiling to identify differentially expressed miRNAs and groups of co-expressed miRNAs, respectively. These steps are followed by functional annotation and interpretation of the results. Ultimately, the results should be shared with the research community. We have previously discussed methods used in miRNA and gene profiling in (Gupta and Schmitz, 2011; Vera *et al.*, 2014).

MiRNAs, or targets of miRNAs identified as being differentially expressed in one condition compared to the other, may give rise to miRNA function. In the following section I describe approaches for inferring miRNA function independent of the experimental design.

Functional annotation and biomarker detection

One way to computationally determine miRNA-phenotype or miRNA-disease associations is through the identification of miRNA biomarkers from expression profiles in treatment or disease condition as compared to a reference or control condition. Biomarkers in the form of miRNA sets that are used for the classification of samples are derived from supervised machine learning approaches such as random forests, support vector machines, Bayes classifiers, k nearest neighbours, or combinations thereof (Ren *et al.*, 2012). For uncharacterized samples, similarities are identified using unsupervised methods including, for example, hierarchical clustering, self-organizing maps, and non-negative matrix factorization (Ren *et al.*, 2012).

Biomarkers present an empirical evidence for miRNA-phenotype or miRNA-disease associations, while a more specific functional characterization of miRNAs can be achieved by additional means. Several approaches have been proposed and include methods for detecting miRNA-phenotype associations and methods for the construction and analysis of network models integrating miRNA regulation (Cascione *et al.*, 2013). A widely used approach to functionally characterize miRNA sets, e.g. differentially expressed miRNAs from expression profiling experiments, is the functional enrichment analysis (a.k.a. gene set enrichment analysis). To this end, miRNA sets are examined for co-occurrences in biological processes, molecular functions and/or biochemical pathways. This is done, for example, by determining associations of predicted and/or validated miRNA targets to gene ontology (GO[8]) terms and/or pathways (e.g. from KEGG[9]). Statistical methods are then used to test if the number of co-occurrences in a functional class exceeds the number observed in a neutral background (Gupta and Schmitz, 2011; Vera *et al.*, 2014). Alternatively, miRNA target sets can be compared to functional gene signatures that are associated to specific phenotypes, pathways, diseases or tissues. A collection of such gene signatures can be found in the Molecular Signatures Database (MSigDB; Subramanian, 2005). In Huang *et al.* (2008) methods and tools for functional enrichment analyses were reviewed and surveyed.

With respect to miRNAs, tools for the identification of miRNA-phenotype associations have been developed which combine miRNA and/or gene expression data with data from target predictions, validated targets and functional enrichment analyses. These tools include miRò (Laganà *et al.*, 2009), FAME (Ulitsky *et al.*, 2010), miRonTop (Brigand *et al.*, 2010), and MAGIA (Sales *et al.*, 2010).

2.1.5 MiRNA web resources

In the last decade the versatile molecule miRNA has been the subject of significant interest, and not only amongst molecular biologists. Countless research activities contributed to a fast growing body of miRNA-related data ranging from sequence and expression to molecular interaction data. In the bioinformatics field, the development of web based repositories or databases is a way to archive such data and also to make it accessible for the research community and the wider public (Schmitz and Wolkenhauer, 2013). Web services, on the other hand, provide online execution of one or multiple computational tasks or automated analyses. Often web services can be accessed programmatically, i.e. from (local) client software or computer scripts.

---

[8] The Gene Ontology Database provides a controlled vocabulary of terms that describes gene product annotation data in terms of biological processes, molecular functions and cellular components (Ashburner *et al.*, 2000).

[9] The KEGG (Kyoto Encyclopedia of Genes and Genomes) database is a collection of manually curated pathway maps representing functionally related molecular interactions (Kanehisa *et al.*, 2008).

In 2003 the first miRNA-related web resource, the 'microRNA Registry' (today known as miRBase database), was released, providing data on published miRNA sequences (Griffiths-Jones, 2004). It has now become an essential resource for all known miRNAs in animals and plants (Kozomara and Griffiths-Jones, 2011; Schmitz and Vearasilp, 2013). miRBase contains experimentally validated miRNA pre-cursor structures, genomic coordinates and sequences of the maturated form of miRNAs. miRBase forms the basis for many other web resources that provide, for example, mRNA target predictions and disease or tissue associations. Other miRNA-related resources include miRNA sequence databases, resources for predicted and experimentally validated targets of miRNAs, functional annotation and mapping databases, resources of miRNA expression and gene regulation, and other web-based tools and services (Schmitz and Wolkenhauer, 2013).

We proposed a system for the classification of miRNA web resources in which we consider all kinds of data collections and web applications related to miRNA research that can be accessed through the internet (Schmitz and Wolkenhauer, 2013). MiRNA web resources can be categorized based on the data type or the technology or computational approach used to generate the data and can be classified as (i) primary, (ii) secondary or (iii) tertiary resources, where primary web resources collect first-hand data (self-collected or generated), secondary resources gather processed data (computational inferences or predictions), and tertiary resources integrate data from several primary and/or secondary resources. Examples for each class are given in Table 2.4 and in (Schmitz and Wolkenhauer, 2013). An extensive list of miRNA related web resources is also provided in Appendix B (Table B.1).

Resource	Description	URL	PMID
Primary miRNA web resources			
miRBase	The miRNA sequence and annotation database	www.mirbase.org	17991681
TransmiR	Manually curated database of TF-miRNA regulations	http://202.38.126.151/hmdd/mirna/tf	19786497
Secondary miRNA web resources			
FAME	Database that provides functional associations of human miRNAs derived with the FAME algorithm	http://acgt.cs.tau.ac.il/fame	20576699
Magia	Interactive web service that integrates miRNA target predictions with mRNA and miRNA expression data analysis	http://gencomp.bio.unipd.it/magia/start	20484379
Tertiary miRNA web resources			
CoGemiR	Collection of information on miRNA genomic location, conservation and expression data	http://cogemir.tigem.it	18837977
miRMaid	Web framework that integrates miRNA data resources and provides an interface for programmable computer access	www.mirmaid.org	20074352

Table 2.4 MiRNA web resources.
This table contains some exemplary web resources categorized in our proposed classification system. An extensive list of currently available miRNA web resources can be found in Appendix B (Table B.1).

In a miRNA related research project, data extraction and integration from established miRNA web resources facilitates a more comprehensive view of the problem at hand by taking into account all available resources and the latest findings of a particular aspect. Obviously, this can only be achieved by wisely selecting reliable and continuously maintained resources. For efficient extraction and seamless integration of miRNA related data, web resources should provide programmatic access, e.g. via a SOAP (Simple Object Access Protocol) or REST (Representational state transfer) interface. Unfortunately, this is not the case for most of the available services. An example of extensive use of web-based miRNA related and other resources is given in Chapter 3, while the design and implementation of a new miRNA web resource is described as part of Chapter 4.

2.2 Systems biology approaches

To better understand basic properties of miRNA mediated gene regulation and the role of miRNAs in large regulatory networks or signalling pathways, it is necessary to employ a systems biology approach. While network models can help to study consequences of deregulated miRNAs for network properties and topology, kinetic models capture concentration dynamics in miRNA regulated pathways. Both these models can be used to study mechanisms of miRNA regulation in the emergence and progression of diseases.

The integration of miRNA regulation, in mathematical models of gene regulatory networks and signalling pathways is making only slow progress (Vera *et al.*, 2013b). The reason may be that most miRNA-target relationships are only predicted and only a small fraction has been experimentally proven so far. However, in the last couple of years the number of validated miRNA-target interactions has seen a steady increase, which is partially due to the recently developed high-throughput methods for miRNA-target identification (HITS-CLIP, Licatalosi *et al.*, 2008; PAR-CLIP; Hafner *et al.*, 2010; and CLASH, Helwak *et al.*, 2013). Another problem is that often it is not clear how miRNAs regulate their target; whether through translation repression or via deadenylation followed by mRNA degradation. Both modes of action have different consequences for target gene products and their RNA messages (Filipowicz *et al.*, 2008). Besides, it is often unclear how the miRNAs themselves are being regulated (Bandyopadhyay and Bhattacharyya, 2010a).

2.2.1 Modelling miRNA target regulation

By modelling miRNA target regulation, one can mimic *in silico*, the mechanisms and dynamics of post-transcriptional gene regulation by miRNAs. The main objective is to simulate quantitative and time dependent consequences on target mRNA and protein concentrations as well as effects on the network level.

To this end, Levine *et al.* (2007) was the first to derive a quantitative model describing local and global properties of miRNA induced target repression which showed that target repression efficiency is influenced by target specific properties (e.g. the number of miRNA binding sites in the 3′ UTR) but also by different cellular conditions that may alter the target processing rate of a miRNA.

A simple model of miRNA target regulation

In the few mathematical models dealing with miRNA target regulation, either simplified phenomenological rate equation models or detailed mechanistic models have been implemented using ODE systems (Levine *et al.*, 2007; Nissan and Parker, 2008; Zinovyev *et al.*, 2013). In the simplest case, a linear dependence of the target degradation on the miRNA level is assumed in addition to its own basal degradation rate:

$$\delta(miR) = \delta_0 + \tilde{\delta} \cdot miR \tag{2.1}$$

Here, δ_0 represents the basal mRNA degradation rate, while $\tilde{\delta}$ represents the rate of change in mRNA concentration due to miRNA regulation (Schmitz, 2013).

In the work of Khanin and Vinciotti (2008) a general Michaelis-Menten function is proposed that considers saturation in miRNA-mediated target degradation.

$$\delta(miR) = \delta_0 + \frac{\tilde{\delta} miR}{k_m + miR} \tag{2.2}$$

In this equation, the Michaelis constant k_m represents the target concentration at which the miRNA-mediated degradation rate is half of the maximally achievable one. Of course, miRNA concentrations are also subject to temporal changes which depend at least on synthesis and degradation parameters, in the following equation represented by s and δ_{miR}, respectively:

$$\frac{d}{dt}miR = s - \delta_{miR}miR \tag{2.3}$$

In a model accounting for the regulation of several targets, the individual target mRNA concentration dynamics, considering transcriptional and post-transcriptional regulation by TFs and miRNAs, can be described as follows:

$$\frac{d}{dt}Tgt_i = k_{syn\_i} - (\delta_{0i} + \frac{\widetilde{\delta_i}miR}{k_i + miR}) \cdot Tgt_i(t) \tag{2.4}$$

This equation includes the basal target transcription rate k_{syn_i}, which itself depends on one or more transcription factors and the basal target degradation rate δ_{0i}, both for target mRNA i, as well as the parameters of the miRNA-mediated target degradation $\widetilde{\delta_i}$ and k_i.

As mentioned earlier, miRNAs can act on their targets in different ways, i.e. by (1) blocking translation initiation, repressing cap recognition or 60S subunit joining, (2) inhibiting translation elongation or causing early translation termination by triggering ribosome drop-off, or facilitating proteolysis of nascent polypeptides; (3) target mRNA degradation caused by destabilization through mRNA deadenylation, or (4) target cleavage in case of perfect sequence complementarity (Filipowicz *et al.*, 2008; Figure 1.4). Therefore, the described simplified model can only account for mechanisms (3) and (4) in which the dynamics of target mRNA concentration depend on miRNA regulation. However, in mechanisms (1) and (2) miRNAs together with the RISC cause changes in target protein concentration, while the total mRNA concentration is in most cases unaffected (Figure 2.5).

One year after Levine's model was published (Levine *et al.*, 2007), Nissan and Parker investigated miRNA regulation in cap-dependent and independent translation initiation by means of mathematical modelling (Nissan and Parker, 2008). They identified different rate-limiting steps in the initiation process and thereby found a plausible explanation for differing experimental observations of miRNA target regulation in the literature. The most detailed mathematical model of miRNA-mediated target regulation so far was introduced by Zinovyev *et al.* (2013) in which the authors included all currently known and hypothesized regulatory mechanisms by which miRNA and the RISC interfere with the target translation process. Their simulation results nicely describe dynamic patterns that correspond to the different regulation mechanisms.

However, miRNAs and their targets are also influenced by transcriptional activators/inhibitors (TFs) and system features like time delays, feedback and feed-forward loops. Only by including these factors into a kinetic model of miRNA target regulation more realistic predictions can be achieved (Schmitz, 2013).

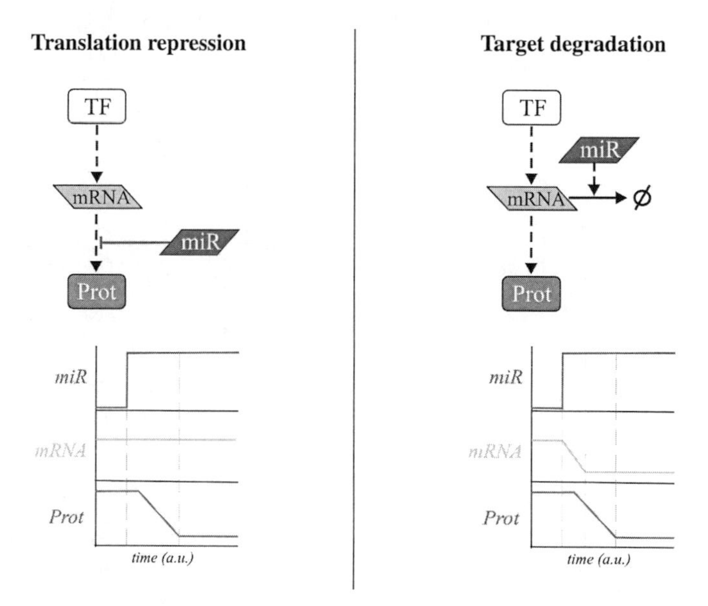

Figure 2.5 Two modes of miRNA induced target repression.
Left: Target translation can be repressed by blocking translation initiation/elongation or by causing early translation termination. These mechanisms lead to a decline in target protein concentration, while the target mRNA concentration is unaffected. **Right:** In this mode the target mRNA is either cleaved (in case of perfect sequence complementarity) or deadenylated, which leads to mRNA destabilization followed by a rapid degradation. These mechanisms cause a drop in target mRNA as well as target protein concentration (a.u.: arbitrary unit).

Modelling target repression efficiency

Functional miRNA target interactions primarily depend on seed site complementarity. Furthermore, target site accessibility, location, hybridization energy and the number of target sites have been proposed as putative features with impact on the effectiveness of miRNA-mediated target regulation (Kertesz *et al.*, 2007; Didiano and Hobert, 2008). However, once a functional miRNA-target relationship has been identified, it is necessary to also determine the target repression efficiency, i.e. the degree to which a miRNA can change the target mRNA and protein concentration. As mentioned before, it is still not clear which mechanisms lead to enhanced or reduced target repression efficiency. Therefore, parameter values for target repression efficiency can only be derived from quantitative time-series data of target concentrations obtained from miRNA transfection experiments.

A common observation, however, is that the target repression efficiency is enhanced when multiple binding sites for the same miRNA exist in the 3′ UTR of the target mRNA. Target repression also increases with the abundance of available miRNAs. The equations (2.1) and (2.2) can be extended by a Hill coefficient ($h \geq 1$) for considering the effect of multiple miRNA binding sites:

$$\delta(miR) = \delta_0 + \tilde{\delta} \cdot miR^h \tag{2.5}$$

$$\delta(miR) = \delta_0 + \frac{\tilde{\delta} \cdot miR^h}{k_m + miR^h} \tag{2.6}$$

Target regulation by multiple miRNAs and miRNA multiplicity

Most previously published mathematical models of miRNA target regulation consider only pairwise miRNA-target interactions. However, it is known that miRNAs often regulate mutual targets in a concerted way and also that certain miRNA species can regulate several targets in parallel (Shalgi *et al.*, 2007; Figure 1.5). The latter, also referred to as target multiplicity, reduces the repression efficiency as per target (Schmitz, 2013). In this case often only a mild reduction in target mRNA and/or protein concentrations can be observed, as for example described in (Selbach *et al.*, 2008). Multiplicity was also considered in the work of Khanin and Vinciotti (2008) who used time-series microarray data of miRNA-124a transfected human HepG2 cells to parameterize a kinetic model of post-transcriptional gene regulation.

Stronger repression, on the other hand, is achieved when a single target is regulated by multiple miRNAs or when multiple miRNA binding sites reside in a target sequence (Selbach *et al.*, 2008a). Furthermore, it was shown by Doench and Sharp (2004) that miRNA pairs with neighbouring target sites can cooperate and thereby achieve an enhanced target repression efficiency. These mechanisms of enhanced miRNA target regulation build the main focus of this thesis and are therefore discussed in detail in the remaining chapters.

2.2.2 MiRNA regulatory network modelling

The awareness of post-transcriptional gene regulation realized by miRNAs is slowly entering the domain of gene regulatory network models. Models integrating transcriptional and post-transcriptional levels are sometimes referred to as composite gene regulatory networks which are supposed to provide a broader picture of the multi-level dynamics of target gene regulation (Schmitz, 2013; Figure 2.6). These network models mimic the interplay between transcription activators/inhibitors, for example in the form of trans-acting genes (a.k.a. TFs) and translation enhancers/inhibitors like miRNAs.

Composite gene expression control is often necessary to realize adaptable gene expression profiles necessary to cope with context specific requirements. In these networks, regulatory motifs such as feedback and feed-forwards loops (FBL and FFL) emerge that integrate miRNAs, TFs and target genes (Zhao *et al.*, 2013b; Vera *et al.*, 2013b). Network motifs are frequent and statistically significant sub-graphs or patterns. A comprehensive review on network motifs and their functions can be found in (Alon, 2007a). Tsang and co-authors found that miRNA-mediated FBLs and FFLs are recurrent motifs in mammalian gene regulatory networks (Tsang *et al.*, 2007). FBLs can be either positive or negative, while FFLs are typically characterized as being coherent or incoherent (see Figure 2.7-Figure 2.10). Depending on the type of loop different time-dependent patterns of target concentrations can emerge (see Figure 2.7-Figure 2.10).

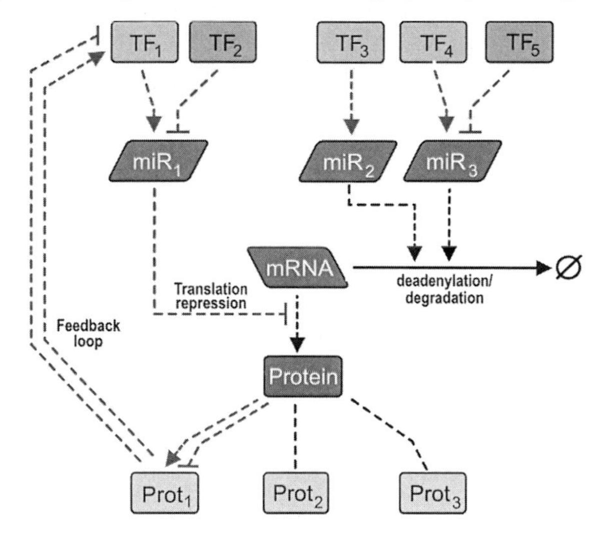

Figure 2.6 Composite regulatory network involving miRNAs, TFs and protein-protein interactions.

Feedback loops

In a positive FBL a biochemical event triggers an upstream process which in turn promotes the event. This mechanism mediates signal amplification and prolonged signal duration. However, as miRNAs negatively regulate their target genes they are typically involved in negative FBLs (Figure 2.7).

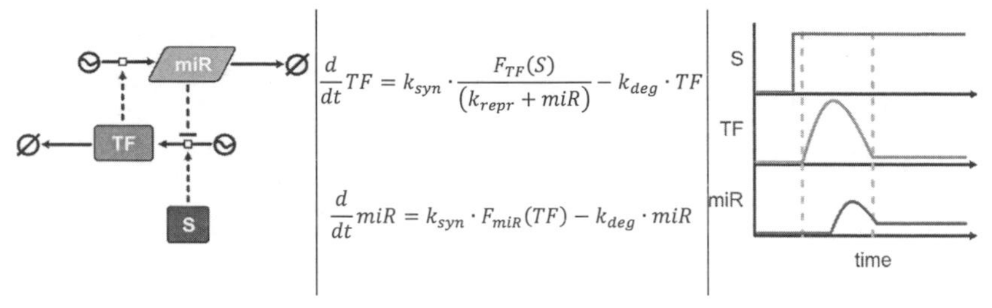

Figure 2.7 Simple negative FBL.
Left: In this network motif miRNA expression is activated by a TF (dashed arrow) which is in turn negatively regulated by its own miRNA target (dashed hammerhead arrow). **Middle**: A simple ODE system describing TF and miRNA concentration dynamics. The TF concentration ($\boldsymbol{TF}$) depends on an external stimulus ($\boldsymbol{S}$), miRNA inhibition and basal degradation. The miRNA concentration ($\boldsymbol{miR}$) depends on TF activation and its own basal degradation. **Right**: The plots illustrate the time-dependent response of TF and miRNA under the influence of an external stimulus. Increased TF activation through an external signal is compensated by the downstream activation of the miRNA that suppresses the TF expression.

In an elementary miRNA induced negative FBL, a miRNA activator, e.g. a TF, is also target of the same miRNA. Negative FBLs typically maintain homeostasis, fine-tune gene expression and protect steady-state concentrations of components in the system against transcriptional noise and uncontrolled fluctuations (Tsang *et al.*, 2007). A transiently increased TF activation, for example, would be compensated by the downstream activation of the miRNA that suppresses the TF expression. In other cases, a negative FBL can mediate periodically oscillating concentrations of components in the loop (Aguda, 2013).

Another commonly observed network motif that includes miRNA regulation is the double negative FBL in which, in the simplest case, a miRNA is negatively regulating its own suppressor (Figure 2.8). Sometimes this regulatory motif is also referred to as coherent feedback loop (Nikolov and Vera, 2013). It can induce bistable expression in both the miRNA and its suppressor which is the characteristic of a bistable toggle switch. In case the suppressor is a TF activated by an external stimulus, the system can turn a transient TF activation into a longer lasting cellular response (Johnston *et al.*, 2005).

miR
TF
S

$$\frac{d}{dt}TF = k_{syn} \cdot \frac{F_{TF}(S)}{(k_{repr} + miR)} - k_{deg} \cdot TF$$

$$\frac{d}{dt}miR = \frac{k_{syn}}{(k_{repr} + TF)} - k_{deg} \cdot miR$$

S
TF
miR
time

Figure 2.8 Double negative FBL.
Left: In this diagram the TF negatively regulates its miRNA target (dashed grey hammerhead arrow) which is in turn negatively regulating its transcriptional suppressor (dashed black hammerhead arrow). **Middle**: These interdependencies can be encoded in two ODEs which describe time-dependent concentration changes of the TF and the miRNA. **Right**: In this motif a transient activation of the TF by an external signal *S* can trigger a permanent transition from miRNA-induced TF repression to TF-mediated miRNA repression.

Feed forward loops

Recurring motifs in gene regulatory networks are FFLs. In most FFLs that involve miRNAs, a miRNA and its TF share a mutual target gene, e.g. the TF is regulating the expression of the miRNA as well as that of the miRNA target (Figure 2.9 & Figure 2.10). In this scenario, both the miRNA and its target are co-expressed due to the co-regulation by the mutual TF. Therefore, it is more likely that the miRNA can act on its target (Aguda, 2013).

From this FFL composition two possible types can emerge: (i) a coherent FFL (Figure 2.9), and an incoherent FFL (Figure 2.10). In the former case a miRNA and its transcriptional activator, both regulate their mutual target gene in a coherent, i.e. negative, manner, while in the latter case they do it in an incoherent way, e.g. the miRNA suppresses the target and a TF activates both the miRNA and their mutual target gene. In a coherent FFL, target repression is secured and prolonged even beyond silencing of the suppressive TF (Figure 2.9). In an incoherent FFL, a step-like TF activation, for example, can induce a pulse response of the target protein level (Figure 2.10).

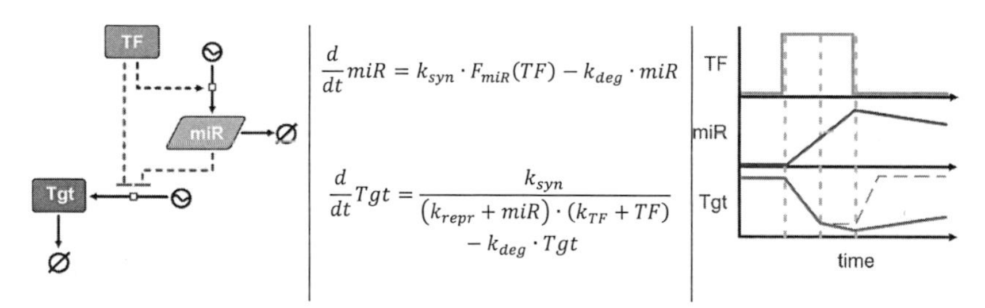

Figure 2.9 Coherent FFL.
Left: In this motif a miRNA and its TF have a mutual target gene which both regulate in a coherent, i.e. negative, manner (dashed hammerhead arrows). **Middle**: The ODE system now contains an equation for the time-dependent change in target concentration which is negatively influenced by the TF and the miRNA. **Right**: Consequences for miRNA and target expression upon a transient TF activation are illustrated in these plots. Without miRNA the target concentration (Tgt) quickly returns to normal after de-repression (dashed line), while in case of miRNA activation the target repression is prolonged (solid line).

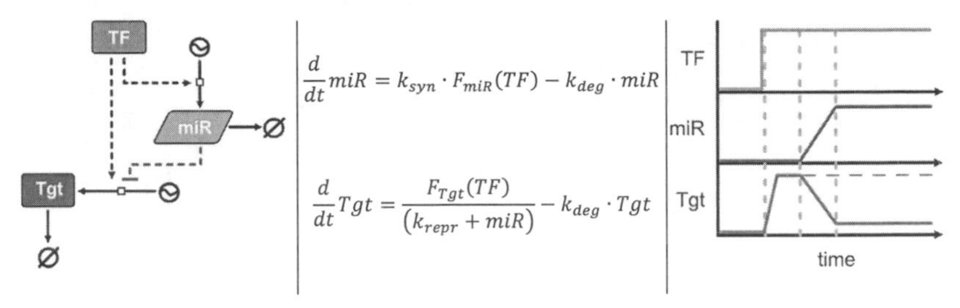

Figure 2.10 Incoherent FFL.
Left: In this motif a miRNA represses target expression (dashed hammerhead arrow) and the TF activates both the miRNA and their mutual target gene (dashed arrows). **Middle**: The ODE for the target concentration dynamics has been modified to account for this change. **Right**: A possible scenario of step-like TF activation is illustrated which has the consequence of a pulse response of the target protein level (solid line). Without miRNA the target activation would be sustained (dashed line).

Re and colleagues (2009) identified 638 putative miRNA-mediated FFLs in human regulatory networks. However, since then the number of known miRNAs has increased five-fold. Therefore, the number of miRNA-mediated FFL will be much higher than previously thought.

Martinez and Walhout summarized in a review the knowledge on validated FBLs and FFLs involving miRNAs and TFs (Martinez and Walhout, 2009). They revealed the reciprocal regulation between these two regulatory molecules and how they coordinate the regulation of shared target genes. Other works have used mathematical modelling to study dynamic properties of miRNA-involving network motifs. For example, Xu *et al.* set up deterministic and stochastic models of miRNA-involving motifs and studied the effect of different input settings on the target gene dynamics (Xu *et al.*, 2009). They concluded that in all cases the systems showed strong robustness of target gene expression against random transcriptional perturbations. The same approach was employed by Osella *et al.* who demonstrated that incoherent FFLs involving miRNAs are efficiently buffering transcriptional noise by dampening fluctuations in target gene expression (Osella *et al.*, 2011). Another interesting feature of miRNA-involving network motifs was shown by Xie and colleagues who have demonstrated that miRNAs can either mediate target gene oscillation or prevent the same (Xie *et al.*, 2007). This is possible when a miRNA is involved in a delayed negative FBL and depends on the severity of its effect on mRNA degradation.

Network biomarkers

MiRNA biomarkers are used to classify uncharacterized samples of miRNA expression based on a classifier previously trained with miRNA expression profiles. MiRNA biomarkers were demonstrated to be very robust and accurate in sample classification and patient stratification. Circulating miRNAs in particular are stable under handling conditions and their presence in all kind of body fluids makes them ideal markers for early prognosis and therapy response monitoring in a variety of diseases (Schwarzenbach *et al.*, 2011). Network biomarkers were recently proposed as an alternative to the classic biomarkers based on miRNA or gene sets. Network biomarkers are models of (signalling) pathways, gene regulatory networks or metabolic networks in the form of molecular interaction graphs (Wu *et al.*, 2012). The difference from 'normal' biomarkers is that network biomarkers integrate heterogeneous data, e.g. data on transcriptional/post-transcriptional gene regulation, transcriptomics/proteomics and molecular interaction data. Dynamic network models (biomarkers) can be constructed through systematic and quantitative data acquisition. Thereby network properties and topology dynamics can be modelled computationally. Such models can be used to simulate network evolution and responses to perturbations (Erler and Linding, 2009). The construction of network models based on heterogeneous data can be facilitated through different computational approaches, e.g. gene regulatory network inference (Hecker *et al.*, 2009), transcription factor binding site (TFBS) prediction (Bulyk, 2003), miRNA-target site prediction (Ritchie *et al.*, 2013), protein-RNA binding-site prediction (Zhao *et al.*, 2013a), prediction of protein-protein interactions (Salwinski and Eisenberg, 2003), molecular network comparison (Sharan and Ideker, 2006) and more (Vera *et al.*, 2014).

2.2.3 Modelling the role of miRNAs in pathogenesis and disease progression

MiRNA-mediated gene regulation is also studied in the context of pathogenesis and disease progression. As mentioned before, causal effects which these small molecules induce in the context of a number of different human pathologies have been validated in e.g. autoimmune diseases, cardio-vascular diseases, neurodegenerative diseases, metabolic diseases and cancer. The latter has also been subject of a number mathematical models involving miRNA regulation. See Schmitz *et al.* (2013) for a comprehensive overview.

In the context of cancer, miRNAs can act as tumour suppressors, oncogenes (a.k.a. *oncomir*) as well as mediators of metastasis (a.k.a. *metastamir*) (Farazi *et al.*, 2013; Bushati and Cohen, 2007; Calin and Croce, 2006). Systems biologists have started to use data-driven mathematical models to elucidate the role of miRNAs in disease emergence and during the course of disease progression. For example, Khanin and Vinciotti (2008) constructed a model based on time-series transcriptomics data that describes target gene expression dynamics based on transfection experiments with the tumour suppressor miR-124a, which is epigenetically silenced in hepatocellular carcinoma and acute lymphoblastic leukemia (Agirre *et al.*, 2009; Furuta *et al.*, 2010). A network model characterizing miR-204 as a tumour suppressor was constructed by Lee *et al.* (2010) and was used to detect 18 gene targets related to tumour progression. Aguda and colleagues constructed a mathematical model of a small regulatory network relevant in cancer which is composed of *MYC*, the *E2F* transcription factor family and the miRNA cluster miR-17-92 (Aguda *et al.*, 2008). The network includes a miRNA-mediated negative FBL and a FBL between *E2F* and *MYC*. They demonstrated through simulations how miR-17-92 induces bistable behaviour in E2F/MYC protein levels and thereby either oncogenic or tumour suppressor function of the miR-17-92 cluster (Aguda *et al.*, 2008; Aguda, 2013). In Vohradsky *et al.* (2010) authors show how miR-124a induces a switch-like behaviour of its target by influencing target decay. In our own recent study we identified, through bioinformatics high-throughput data analysis, a set of miRNAs involved in the progression of the malignant melanoma (Bhattacharya *et al.*, 2015). One of these miRNAs, miR-638, negatively regulates expression of the oncogene *TFAP2A*, a transcription factor whose expression positively correlates with poor prognosis and advanced stages of the disease (Bar-Eli, 2001; Berger *et al.*, 2005). We showed that epigenetic deregulation of miR-638 via promoter methylation has the potential to suppress apoptosis through repression of *TFAP2A*. To better understand this phenomenon of miR-638 mediated regulation of apoptosis we integrated knowledge from publications and databases into a regulatory network from which we derived an ODE-based mathematical model following the same approach as described in Chapter 3 of this thesis. The network involves a double negative feedback loop (TFAP2A ⊣ miR-638 and miR-638 ⊣ TFAP2A). The model simulations indicate that methylation of the TFAP2A binding site in the miR-638 promoter can trigger a permanent transition from TFAP2A-mediated miR-638 repression to miR-638-induced TFAP2A repression. This suggests that TFAP2-induced expression of pro-apoptotic genes is stopped and cancer cells can evade apoptosis initiation.

Several of the bioinformatics and systems biology methods discussed in this chapter are integrated and enhanced in the following two chapters for studying the complex regulatory network of miRNA target hubs (Chapter 3) and for predicting targets of cooperative miRNA regulation (Chapter 4).

3 Reconstruction and analysis of networks regulating miRNA target hubs

This chapter is based on the following publications:

Nikolov S, Vera J, **Schmitz U**, Wolkenhauer O (2010) A model-based strategy to investigate the role of microRNA regulation in cancer signalling networks. Theory Biosci 130: 55-69

Lai X*, **Schmitz U***, Gupta SK, Bhattacharya A, Kunz M, Wolkenhauer O, Vera J (2012). Computational analysis of target hub gene repression regulated by multiple and cooperative miRNAs. Nucleic Acids Res 40 (18), 8818–8834.

Lai X, Bhattacharya A, **Schmitz U**, Kunz M, Vera J, Wolkenhauer O (2013). A systems biology approach to study microRNA-mediated gene regulatory networks. BioMed Res Int 2013; 2013:703849.

Vera J, Lai X, **Schmitz U**, Wolkenhauer O (2013). MicroRNA-Regulated Networks: The Perfect Storm for Classical Molecular Biology, the Ideal Scenario for Systems Biology. Adv Exp Med Biol, 774, 55-76.

(The '*' symbol denotes equal contributions to the respective work.)

Synopsis

*MiRNA target hubs are genes that can be regulated, at the post-transcriptional level, by a comparatively large number of miRNA species. Together with the transcriptional regulation of these genes by TFs, large regulatory networks arise from which complex and process-specific regulatory patterns emerge that facilitate a fine-tuned adjusting of target hub expression. In this chapter I describe an integrative workflow for the reConstruction and Analysis of networks regulating miRNA Target Hubs (**CATH**). The workflow integrates bioinformatics and systems biology methods for achieving a better understanding of the complex dynamics involved in miRNA target hub regulation. Furthermore, I describe a case study in which we implemented and tested the CATH workflow for the case of the miRNA target hub CDKN1A (a.k.a. p21/Cip/Waf1). The results of this work provide evidence that miRNA target hubs are subject to fine-tuned transient and long-term post-transcriptional gene regulation that is adjusted according to the needs in different cellular scenarios and that collective and cooperative regulation by miRNAs is useful, e.g. for buffering transcriptional noise. This chapter demonstrates that the CATH workflow provides a comprehensive means for investigating mechanisms of collective miRNA repression.*

3.2 Motivation for and design of the CATH workflow

In the previous chapter I introduced bioinformatics and systems biology methods for the investigation of miRNA target regulation. In this chapter many of these methods are implemented in a comprehensive workflow (CATH) for the analysis of miRNA target hub regulation.

It is currently known, that miRNAs form an integral part of gene regulation at the post-transcriptional level. They are involved and can modulate various cell functions by regulating critical target genes. This adds an additional layer to the already complex regulatory networks composed of genes, proteins and small molecules that control cellular functions (Bartel, 2009). To make things worse, in terms of complexity, most miRNAs can regulate a large number of target genes while gene transcripts are often the target of multiple miRNA species (Selbach *et al.*, 2008; Shalgi *et al.*, 2007; Hon and Zhang, 2007). To better understand gene regulation by multiple miRNAs we investigated the phenomenon of miRNA-mediated repression of target hub genes. Target hubs are in general subject to regulatory control by many trans-acting elements[10]. The term, target hub gene, was previously used in (Borneman *et al.*, 2006) to describe genes that are transcriptionally controlled by several TFs. In the context of miRNA-target regulation, this term was first used by Shalgi *et al.* (2007) who stated that a gene regulated by at least 15 different miRNAs can be referred to as miRNA target hub. This class of targets is particularly interesting because many of these hubs are transcriptional regulators and they are often involved in various developmental processes (Shalgi *et al.*, 2007). However, in our investigations we considered not only collective miRNA regulation of target hub genes but also other regulatory elements involved in the modulation of target hub expression. To this end, we integrated published data and computational predictions of interactions between TFs and the target gene, TFs and miRNAs, as well as target protein-protein interactions. A generic illustration of a target hub regulatory network is presented in Figure 3.1.

[10] Trans-acting or trans-regulatory elements are genes that give rise to TFs, which can regulate the transcription of other genes.

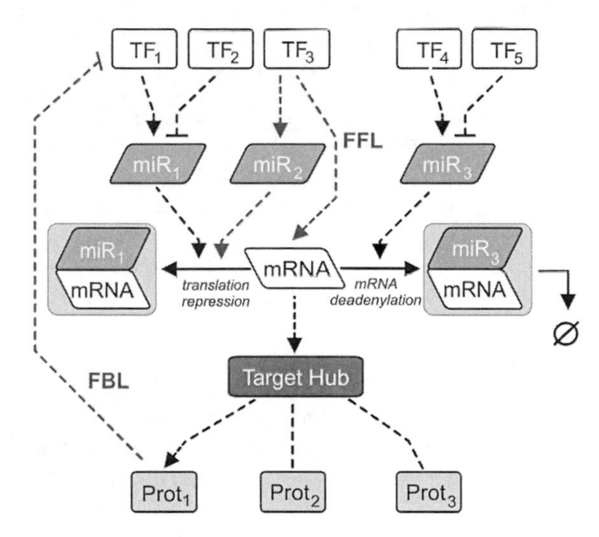

Figure 3.1 Regulatory network of a miRNA target hub.
For the systemic analysis of miRNA target hub regulation it is important not only to consider miRNA-target interactions but also TF-miRNA, TF-gene and protein-protein interactions. Thus all direct and indirect trans-acting regulators on the transcriptional, post-transcriptional and post-translational levels are incorporated into the analysis. TFs can promote or inhibit transcription of target genes and miRNAs, here indicated by the dashed lines with arrow-heads and hammer-heads respectively, whereas miRNAs can either induce translation repression or mRNA deadenylation. Both mechanisms lead to a reduction in target protein expression levels. These kinds of regulatory networks exhibit interlocked network motifs like feed-forward loops, and also feedback loops when involving protein-protein interactions (the loops are indicated by the red dashed arrows). These motifs may lead to non-linear dynamics in the regulatory control of miRNA target hubs.

The first experimentally fully supported miRNA target hub is a well-known cell-cycle regulator and tumour suppressor, the cyclin-dependent kinase inhibitor 1A (*CDKN1A*) which is also known under the synonyms *p21*, *CIP1* or *WAF1* (among others). It acts as inhibitor of G_1 cyclin-dependent kinases, thereby mediating cell cycle arrest in response to stress signals, e.g. DNA damage induced stress (Harper *et al.*, 1993; Gartel and Tyner, 1999). In a large-scale *in vitro* analysis performed by Wu and colleagues, 266 miRNAs were tested for their *CDKN1A* regulation efficacy using luciferase assays (Wu *et al.*, 2010). These miRNAs represent the joint results of predictions from different computational algorithms that follow different but complementary approaches. The algorithms used are miRanda, PicTar, RNA22 and TargetScan (Enright *et al.*, 2003; Krek *et al.*, 2005; Miranda *et al.*, 2006; Lewis *et al.*, 2005). MiRNA target prediction algorithms and experimental validation of miRNA-target interactions are discussed in Chapter 2 as well as in (Bhattacharya and Kunz, 2013). A subset of 28 miRNAs induced a significant reduction in luciferase activity. These results were underpinned by Western blot analyses of CDKN1A protein levels after miRNA transfection in HEK 293 cells (Wu *et al.*, 2010). This was to our knowledge the first case in which a miRNA target hub was experimentally validated in a controlled experimental setting.

However, by solely considering the individual miRNA-target interactions one cannot infer target hub expression profiles for different contexts, i.e. cellular scenarios, as other factors may concurrently influence the target hub expression. The goal of our investigation was to establish a workflow suitable for the in depth analysis of the complex mechanisms by which a sophisticated target hub gene regulation is mediated by multiple miRNAs and other trans-acting elements. The CATH workflow that we designed and implemented integrates bioinformatics and systems biology approaches and consists of seven steps: (i) Data integration: the retrieval and integration of molecular interaction data from the literature, public databases but also predicted interactions to construct a regulatory network for a target hub of interest, (ii) Scoring: the assessment of molecular interactions and assignment of confidence scores, (iii) Network reconstruction: standardized visualization of molecular interactions in a network graph, (iv) Network analysis: for deriving topological properties and network motifs, (v) Model derivation: the derivation of a kinetic model from the regulatory network that describes the dynamics of the molecules involved, (vi) Predictive simulations: for example predictions of target hub expression in different cellular scenarios, and (vii) Validation: model-driven experiments in order to validate predictions from the model simulations. See Figure 3.2 for an illustration of the CATH workflow. Each step is discussed in detail below.

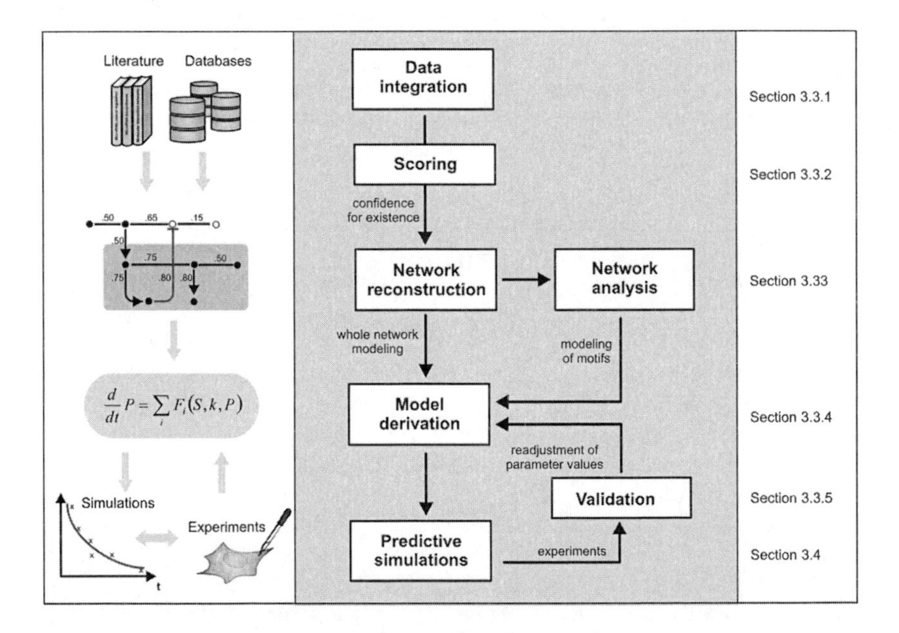

Figure 3.2 Schematic representation of the CATH workflow.
Data of regulatory interactions from the literature and public repositories are integrated into a network representation, which includes confidence scores for each molecular interaction. The network is analysed with respect to structure and topology and network motifs (e.g. FFLs and FBLs) are identified. From the network, a mathematical model is derived and parameterized using quantitative data from experiments and other resources. The model is then calibrated in iterative cycles of model predictions and experimental validation. The calibrated model (predictive model) describes miRNA dependent target hub regulation in signalling pathways or gene regulatory networks and can be used to test hypotheses and simulate complex biological scenarios associated with cellular function-related variability.

We implemented the CATH workflow in a case study to investigate the sophisticated cellular context-dependent regulation of the miRNA target hub *CDKN1A*. We analysed the structure and dynamics of the literature and database-derived *CDKN1A* regulatory network using methods of network analysis and an inferred kinetic model composed of ordinary differential equations. Thereby, we identified putative housekeeping[11] miRNAs which are regulated by a set of TFs that are involved in many distinct cellular functions. Furthermore, we included in our model the recently discovered phenomenon of cooperative target regulation by pairs of miRNAs (Doench and Sharp, 2004b) and investigated possible consequences for the efficiency and dynamics of target repression. We computationally identified several pairs of miRNAs that may cooperate in the repression of *CDKN1A* and experimentally validated the cooperativity between two of them (miR-572 and miR-93). We further used these two miRNAs as an example to investigate three hypothetical target regulation mechanisms by multiple miRNAs. We found that one miRNA is sufficient to completely repress a target when highly expressed. However, the same effect can be achieved by two cooperating miRNAs, both of which are only expressed on a medium level. Furthermore, in a series of simulation experiments we predicted the CDKN1A protein expression for nine different cellular functions. Interestingly, we found that our simulation results are well in agreement with the respective CDKN1A expression profiles described in the literature, particularly those where the CDKN1A expression is critical for the cellular function, as for example in the cellular response to DNA damage or in apoptosis.

3.3 Research strategy

3.3.1 Data integration

For the reconstruction of a regulatory network of a miRNA target hub data from various sources have to be integrated (literature, databases or experimental data). It is important, however, to take care of the timeliness and reliability of the data, the notation used and

[11] Housekeeping miRNAs just as housekeeping genes are elementary cell components that are required for the maintenance of basic cellular function. They are expressed under normal conditions in most cell types. Therefore some of them are used as endogenous control in miRNA profiling experiments.

the accuracy of the underlying computational algorithms (Schmitz and Wolkenhauer, 2013). Furthermore, it is often necessary to use different resources which may provide complementary information.

In target hub regulatory networks we consider four different kinds of molecular interactions: (i) TF-target gene interactions; (ii) TF-miRNA interactions; (iii) miRNA-target interactions; and (iv) target-protein interactions. For each type of interaction a number of public databases exist that provide these kinds of data (Table 3.1). For our *CDKN1A* case study, we integrated information from the following resources:

Type of interaction	Open resources	URL
Protein-protein interactions	HPRD	www.hprd.org
	STRING	http://string-db.org
TF-gene interactions	Literature	
	UCSC browser	https://genome.ucsc.edu
MiRNA-target interactions	miRecords	http://c1.accurascience.com/miRecords
	TarBase	http://diana.cslab.ece.ntua.gr/tarbase
	miRTarBase	mirtarbase.mbc.nctu.edu.tw
	miRWalk	www.umm.uni-heidelberg.de/apps/zmf/mirwalk/
	miRGen	diana.cslab.ece.ntua.gr/mirgen
TF-miRNA interactions	TransmiR	http://www.cuilab.cn/transmir
	UCSC browser	https://genome.ucsc.edu

Table 3.1 Resources for reconstructing regulatory interaction networks.

Protein-protein interaction data were retrieved from the Human Protein Reference Database (HPRD, release 9.0; Keshava Prasad *et al.*, 2009) and the STRING database (release 9.0; Szklarczyk *et al.*, 2011). For TFs that regulate target gene expression, the most exhaustive and reliable source was the scientific literature. However, we complemented the list of validated TF-*CDKN1A* interactions with putative TFs that are associated with conserved transcription factor binding sites (human, mouse, rat) located in the 5kb upstream region of the *CDKN1A* gene. This information was extracted from the table of TFs with conserved binding sites (tfbsConsFactor, tfbsConsSite, hg18) in the UCSC genome browser (Karolchik *et al.*, 2003). A complete list of putative and validated TF-*CDKN1A* relationships can be found in the Appendix (Table C.1). MiRNA-target relationships can be extracted from databases of validated miRNA-target interactions like the miRecords, TarBase and miRTarBase which are resources of literature derived and manually curated data (Xiao *et al.*, 2009; Sethupathy *et al.*, 2006b; Hsu *et al.*, 2011).

Moreover, there exists a large number of resources on miRNA-target interactions based on different prediction algorithms, like for example the miRWalk database or the miRGen database which provide a collections of miRNA-target predictions (Dweep *et al.*, 2011; Alexiou *et al.*, 2010). We reviewed these and other miRNA-related resources in (Schmitz and Wolkenhauer, 2013). However, as mentioned before, we solely considered miRNA-*CDKN1A* interactions that were experimentally validated in the paper by Wu *et al.* (2010). Their specific binding sites, however, were derived from the microRNA.org web resource (Betel *et al.*, 2008) representing the results of the miRanda target prediction algorithm (Enright *et al.*, 2003). Data of TFs that regulate the expression of miRNAs involved in the *CDKN1A* regulatory network were extracted from the literature-based and the manually curated database TransmiR (release 1.0), which provides data on TF-miRNA interactions with experimental support (Wang *et al.*, 2010). However, because validated TFs are known only for a few miRNAs we also considered TFs with predicted binding sites in the promoter regions (10 kb upstream region[12]) of *CDKN1A* regulating miRNAs. Several sources which are based on different algorithms exist. We considered only those TF-miRNA interactions that were predicted in two out three of the following resources: (i) PuTmiR (release 1.0; Bandyopadhyay and Bhattacharyya, 2010b); (ii) MIR@NT@N (version 1.2.1; Le Béchec *et al.*, 2011); and (iii) from the table of TFs with conserved binding sites (tfbsConsFactor) in the UCSC genome browser (hg18; Karolchik *et al.*, 2003).

3.3.2 Confidence scores for molecular interaction data

The reliability of molecular interactions given in a regulatory network can be indicated by a confidence score. However, this is not yet common practice. Confidence scores can be found in some resources of protein-protein interaction data, like STRING, MINT or IntAct (Szklarczyk *et al.*, 2011; Ceol *et al.*, 2010; Kerrien *et al.*, 2012). Nevertheless, to date no confidence scoring systems have been proposed for a miRNA regulatory network and the molecular interactions involved.

In our case study, in order to provide a sense of how reliable the information on molecular interactions in the *CDKN1A* regulatory network is, we designed a scoring system and assigned confidence scores to each edge (interaction) in the network. We used the system referred to as MIscore (http://code.google.com/p/miscore) implemented in the IntAct database (Kerrien *et al.*, 2012) as a template to derive a more systemic system (beyond protein-protein interactions). For each type of molecular interaction in the miRNA regulatory network we derived a customized scoring scheme based on manifestations in the literature, experimental techniques employed, predictions and more. The system assigns scores in a range between 0 and 1 to interactions that are at one extreme totally uncertain and at the other most reliable. Each scoring scheme makes use of general and

[12] We selected this region following the method in Shalgi *et al.* (2007), where the 10 kb upstream region is indicated as 'putative regulatory region of miRNAs'.

specialized scoring criteria. One general scoring criterion is the number of scientific publications with experimental evidence for a molecular interaction $S_P(n)$. The *publication score* is calculated as follows:

$$S_P = log_{(b+1)}(n+1) \tag{3.1}$$

In this equation n represents the number of scientific publications reporting a molecular interaction, while b is the number required for assigning the maximum possible score ($S_P = 1$). The value for b depends on the type of molecular interaction with respect to the coverage in scientific publications. To define b and other weighting factors used for our confidence scoring system we conducted a survey among experts in the field. More information on the survey, the participating laboratories and the results of the questionnaire can be found in Appendix C.1.

Protein-protein interactions

The confidence scores assigned to protein-protein interactions (PPI) are based on two criteria: (i) the number of scientific publications with experimental evidence for a PPI, and (ii) the interaction type. In the equation for the PPI score:

$$S_{PPI} = \frac{k_P \cdot S_P(n) + k_T \cdot S_T}{k_P + k_T} \tag{3.2}$$

the weight k_P assigned to the score for the number of supportive publications $S_P(n)$ was 0.7 in the case of the *CDKN1A* regulatory network, while more weight was assigned to the type of interaction ($k_T = 0.9$). $S_P(n)$ is calculated according to equation (3.1), while the maximum possible score is achieved in the case of six or more supportive publications ($b = 6$). The score for the interaction type S_T differentiates between direct ($S_T = 1$) and indirect ($S_T = 0.5$) interactions. In the *CDKN1A* regulatory network, to limit the complexity, we considered direct interactions only.

Transcription factor target gene interactions

In the case of TF-target gene interactions, we differentiate between experimentally supported and predicted interactions. In the *CDKN1A* case study, for example, we extracted the information about confirmed TF-*CDKN1A* interactions from the literature. The confidence score for these interactions was computed based on the following equation:

$$S_{TF} = \frac{k_P \cdot S_P(n) + k_R \cdot S_R + k_{BS} \cdot S_{BS}(m)}{k_P + k_R + k_{BS}} \tag{3.3}$$

This equation again contains a score for the number of publications confirming an interaction $S_P(n)$. This score is computed based on equation (3.1). Another criterion for

TF-target gene interactions that we considered is the type of regulation represented by the score S_R. Here, we differentiate between: (i) direct regulation through TF binding to the target promoter region ($S_R = 1$); and (ii) indirect target regulation, e.g. by complex formation or through a co-factor ($S_R = 0.5$). However, in our *CDKN1A* case study we considered only direct regulation determined through methods for detecting protein-DNA interactions[13]. The third factor that we consider for determining the confidence of a TF-target gene interaction is the number of TF binding sites in the target promoter region represented by the score $S_{BS}(m)$. The *binding sites score* is calculated as follows:

$$S_{BS}(m) = log_{(b+1)}(m+1) \tag{4}$$

In this equation *m* represents the number of binding sites reported or predicted for a TF in the target gene promoter region, while b is the number required for assigning the maximum possible score ($S_{BS} = 1$). In case of the *CDKN1A* network, for m ≥ b = 6 the *binding sites score* S_{BS} equals 1. The weights assigned to the scores for publications, regulation type and number of binding sites in our *CDKN1A* case study are $k_P = 0.7$, $k_R = 0.9$ and $k_{BS} = 0.9$ based on the results of our survey.

Apart from validated TF-target gene interactions, we also considered computationally predicted interactions. The following equation is used to derive the score for predicted TF-target gene interactions:

$$S_{TF} = \frac{k_{Cons} \cdot S_{Cons} + k_{BS} \cdot S_{BS}(m)}{k_{Pred} + k_{BS}} \tag{3.5}$$

Predicted TF-target gene interactions are scored based on the normalized *conservation score* S_{Cons} derived from the UCSC table browser and the *binding sites score* S_{BS}. Furthermore, we introduced a weighting factor ($k_{Pred} = 0.3$) to discriminate predicted interactions from validated ones.

MiRNA-target gene interactions

Following a similar principle as for the other interaction types, in the case of miRNA-target gene interactions the assigned score is composed of scores for the number of scientific publications supporting an interaction, the experimental method used to validate it and the number of (predicted) binding sites of a miRNA in the mRNA 3′ UTR of its designated target. The following equation is used to derive the score for miRNA-target gene interactions:

---

[13] Experimental methods for detecting protein-DNA interactions are: (i) chromatin immunoprecipitation (ChIP) assays; (ii) DNA electrophoretic mobility shift assay (EMSA); (iii) DNA pull-down assay; (iv) microplate capture and detection assay; and (v) reporter assays.

$$S_{miR} = \frac{k_P \cdot S_P(n) + k_M \cdot S_M(method) + k_{BS} \cdot S_{BS}(bs)}{k_P + k_M + k_{BS}} \quad (3.6)$$

In this case the highest weight and therefore the strongest confidence is assigned to the method score ($k_M = 1$), followed by the publication score ($k_P = 0.7$) and the binding sites score ($k_{BS} = 0.3$). The maximal binding sites score S_{BS} is achieved for three or more predicted binding sites. Scores for the experimental methods used to validate a miRNA-target interaction were also established as part of our expert survey. They are supposed to reflect the capability of a method to identify true miRNA-target interactions. The following scores were defined:

Method	Reporter assay	Western blot	qRT-PCR	Microarray
Score S_M	0.8	0.7	0.4	0.3

TF-miRNA interactions

This type of interaction is in principal the same as TF-target gene interactions; therefore we adopted a similar scoring scheme. However, in our *CDKN1A* case study we experienced a lack of experimentally supported TF-miRNA interactions. For that reason we also incorporated predicted interactions. Therefore, the scoring scheme is composed of a *publications score* (S_P) and a *predictions score* (S_{Pred}), which considers the number of algorithms predicting an interaction as well as the number of predicted binding sites. The TF-miRNA confidence score is computed as follows:

$$S_{TFmiR} = \frac{k_P \cdot S_P(n) + k_{Pred} \cdot S_{Pred}(m, bs)}{k_P + k_{Pred}} \quad (3.7)$$

The weights assigned to the *publications score* (S_P) and *predictions score* (S_{Pred}) are $k_P = 0.8$ and $k_{Pred} = 0.5$ respectively. The equation for the determination of the *predictions score* is as follows:

$$S_{Pred}(m, bs) = \frac{k_m \cdot log_{(m_{max}+1)}(m+1) + k_{BS} \cdot log_{(bs_{max+1})}(bs+1)}{k_m + k_{BS}} \quad (3.8)$$

The equation incorporates the number of algorithms that predict an interaction (m) together with the maximum possible number ($m_{max} = 3$) and the number of predicted binding sites (BS) together with a cutoff value defined for the number of binding sites required for achieving the maximal efficiency in the target regulation ($bs_{max} = 6$). For the number of algorithms predicting a TF-miRNA pair and the number of binding sites predicted we assigned the weights $k_m = 1$ and $k_{BS} = 0.7$, respectively.

3.3.3 Network reconstruction and analysis

A molecular interaction network can be implemented as a directed or undirected graph (Barabási and Oltvai, 2004). A gene regulatory network in particular has to be directed to indicate which molecule regulates another. Typically, it is also indicated whether a regulatory relationship is positive or negative to denote for an activation process or inhibition process, respectively (Alon, 2007b). Furthermore, in cyclic graphs one can identify network motifs (recurring sub-graphs or patterns) and regulatory loops such as feedback and feed forward loops (Ideker *et al.*, 2002). For more information on modelling network motifs involving miRNAs see Chapter 2.

It is advantageous to implement a network diagram in a standardized form to avoid ambiguous representations which may be misinterpreted by other researchers. The widely accepted standard for the exchange of molecular interaction networks is the Systems Biology Graphical Notation (SBGN; Le Novère *et al.*, 2009). Several software tools exist to visualize molecular interaction networks in a SBGN complaint fashion, the most popular of which is CellDesigner™ (Funahashi *et al.*, 2003).

The analysis of gene regulatory networks, involving or not involving miRNA regulation and other types of networks, includes the assessment of network parameters and topological properties. This can be done most efficiently in the network visualization software Cytoscape. This open source platform comes with a library of tailored plugins (now referred to as Apps) for integrating network data with, e.g. annotations or gene expression profiles, for analysing network properties and for the identification of network motifs (Shannon, 2003). Despite general parameters like, e.g., the number of nodes and the number of edges, topological properties include the *clustering coefficient*, *network diameter* and *radius*, *average path length*, *average connectivity of nodes,* and more. To determine if a network is robust against perturbations a frequently used method, first suggested by Barabási and Oltvai (2004), is to fit a power law function of the form $y = a \cdot x^{-b}$ to the in- and out- degree distribution of the network vertices (i.e. nodes). In this equation y denotes the number of vertices and x the degree. Despite general network properties and characteristics one can also analyse metrics of network components, i.e. molecule species, like for example their degree or different centrality measures. These metrics help to identify the most influential nodes in a network (Peng and Schork, 2014).

Another aspect which is important in the context of gene regulatory networks is the identification of network motifs. Certain motifs can give rise to a non-linear, and therefore non-intuitive, dynamical behaviour of constituents of such motifs. Network motifs like feedback loops and feed-forward loops involving miRNAs, TFs and other signalling proteins have been described in the literature (Shalgi *et al.*, 2007; Osella *et al.*, 2011; Tsang *et al.*, 2007). For example, when a miRNA inhibits its own TF it is considered a feedback loop and when, for example, a miRNA and its TF share a mutual target gene it is considered a feed-forward loop (Vera *et al.*, 2013b). Again, a detailed analysis of possible loops involving miRNA regulation and their characteristic dynamics is provided in Chapter 2.

Among the many tools that exist for the identification of regulatory loops and for the analysis of network properties we choose the Cytoscape App named NetworkAnalyser (Assenov *et al.*, 2007) for its reliability and its potential to seamlessly integrate into the CATH workflow.

3.3.4 Derivation of a kinetic model

In the systems biology approach it is common to analyse a biological system by first structuring and integrating all available information into a network representation. Thereafter, this network is translated into a mathematical model used to describe and analyse the inherent and possibly induced dynamics of the biological system under investigation. For the analysis of miRNA target hub regulation and based on the available data we chose a kinetic modelling approach based on ODEs. Kinetic modelling can be used to describe the temporal evolution of, e.g. miRNA, mRNA and protein expression levels in a regulatory network. Therefore, to describe the dynamics in the regulation of a miRNA target hub we propose the following set of ODEs:

$$\frac{d}{dt}th_m = k_{syn\_th_m} \cdot f_{act}(TF_{th_m}) - th_m \cdot \left(k_{deg\_th_m} + \sum_i k_{ass\_miR_i} \cdot miR_i \right)$$

$$\frac{d}{dt}miR_i = k_{syn\_miR_i} \cdot f_{act}(TF_{miR_i}) - miR_i \cdot k_{deg\_miR_i} - k_{ass\_miR_i} \cdot th_m \cdot miR_i$$

$$\frac{d}{dt}[th_m|miR_i] = k_{ass\_miR_i} \cdot th_m \cdot miR_i - k_{deg\_comp_i} \cdot [th_m|miR_i]$$

$$\frac{d}{dt}th_p = k_{syn\_th_p} \cdot th_m - k_{deg\_th_p} \cdot th_p$$

The equations account for the time dependent changes in concentration of the target hub mRNA (th_m) and protein (th_p) as well as n regulatory miRNAs (miR_i, $i = 1 \dots n$). Furthermore, one equation is dedicated to the complexes ($[th_m|miR_i]$) that are formed when a miRNA binds to the mRNA target. Processes considered for the target hub mRNA (th_m) are described using the following model parameters: (i) $k_{syn\_th_m}$ for the basal mRNA synthesis mediated by a transcription factor $f_{act}(TF_{th_m})$, (ii) $k_{deg\_th_m}$ for basal degradation; and (iii) $k_{ass_{miR\,i}}$ for the association with a miRNA i. Processes considered for the regulatory miRNAs (miR_i) are: (i) the basal synthesis described by the parameter $k_{syn\_miR_i}$ and the function for the TF activity $f_{act}(TF_{miR_i})$, (ii) the basal degradation described by the parameter $k_{deg\_miR_i}$, and (iii) the association with the target mRNA ($k_{ass\_miR_i}$). Dynamics of each complex ($[th_m|miR_i]$) composed of a miRNA and the target mRNA are described by the parameters: (i) $k_{ass\_miR_i}$ for the complex' assimilation, and (ii) $k_{deg\_comp_i}$ for the miRNA induced degradation. Finally, the processes considered for the target protein (th_p) are: (i) the mRNA mediated synthesis described by the parameter $k_{syn\_th_p}$, and (ii) the protein degradation ($k_{deg\_th_p}$).

The total measurable amount of target mRNA (th_{m_tot}) and each miRNA (miR_{i_tot}) is determined as follows:

$$th_{m\_tot} = th_m + \sum_i [th_m|miR_i]$$

$$miR_{i\_tot} = miR_i + \sum_i [th_m|miR_i]$$

Model structures like this can be implemented in simulation software systems. The frequently used systems are Matlab® and Copasi (Hoops *et al.*, 2006). For our case study, we used the technical computing software Matlab® because of its rich selection of analytical tools. Parameter estimation, for example, was performed using the SBtoolbox2 for Matlab® (Schmidt and Jirstrand, 2006). More details on the construction and parameterization of the ODE model for our *CDKN1A* case study are described in Appendix C.2, while the model simulations are shown and discussed in the section 3.4.5 of this chapter and in Appendix C.3.

3.3.5 Experimental validation

There is a variety of different experimental methods used for investigations in miRNA biology. Some of them are introduced in Chapter 2 of this thesis with a special focus on the validation of miRNA-target interactions. For the *CDKN1A* case study our collaborators from the University of Leipzig validated our predictions associated with the phenomenon of cooperative target gene regulation by pairs of miRNAs. More precisely, cells belonging to a melanoma cell line (SK-Mel-147) were transfected with mimics of two miRNAs that regulate the expression of *CDKN1A* (miR-572 and miR-93). In two separate experiments each miRNA mimic was transfected individually at a concentration of 100 nM and in a third experiment both miRNA mimics were transfected together at 50 nM respectively. The expression of the target gene *CDKN1A* was triggered with 250 nM (pulse-treatment) of doxorubicin, a genotoxic substance inducing DNA damage. Protein lysates were prepared for western blotting at time points 0, 2, 4, 5, 8 and 24hr post treatment. The concentration of CDKN1A was measured by western blotting for each time point. More details on the exact experimental procedure are given in Appendix C.3, while the results are discussed in the following section.

3.4 Case study

3.4.1 Molecular interaction network displays sophisticated *CDKN1A* regulation

As described in the previous section we extracted and integrated data on *CDKN1A* regulation from the literature and public databases that describe TF-*CDKN1A* interactions, miRNA-*CDKN1A* interactions, TF-miRNA interactions (miRNAs regulating *CDKN1A*), and protein interactions of the CDKN1A protein. An Excel table with annotated molecular interaction data and confidence scores can be accessed from the Nucleic Acid Research journal website (http://tinyurl.com/pb5jzmg), a reduced version of which can be found in Appendix C.1 (Table C.1). The annotations include accession numbers, GO terms, PubMed identifiers and other details of the protein interaction partners, miRNAs and TFs. Confidence measures were established based on the scoring system introduced in Section 3.3.2 on the Research strategy. This system integrates weighted scores for publications reporting an interaction, experimental method(s) used, interaction type and computational predictions. Based on the derived data we constructed a regulatory network (Figure 3.3) using CellDesigner™, a graphical editor for drawing biochemical and gene-regulatory networks (Funahashi *et al.*, 2003). CellDesigner™ supports a standardized graphical representation in SBGN (Le Novère *et al.*, 2009) and is SBML compliant, i.e. it saves model wrappers for biochemical reaction networks using the Systems Biology Markup Language (SBML; Hucka *et al.*, 2003), a standard format for sharing models of biochemical and gene-regulatory networks. The annotated SBML file of the regulatory network is available at www.sbi.uni-rostock.de/resources/software/target-hub. A PDF version (vector graphics) of the regulatory network is available under the same URL.

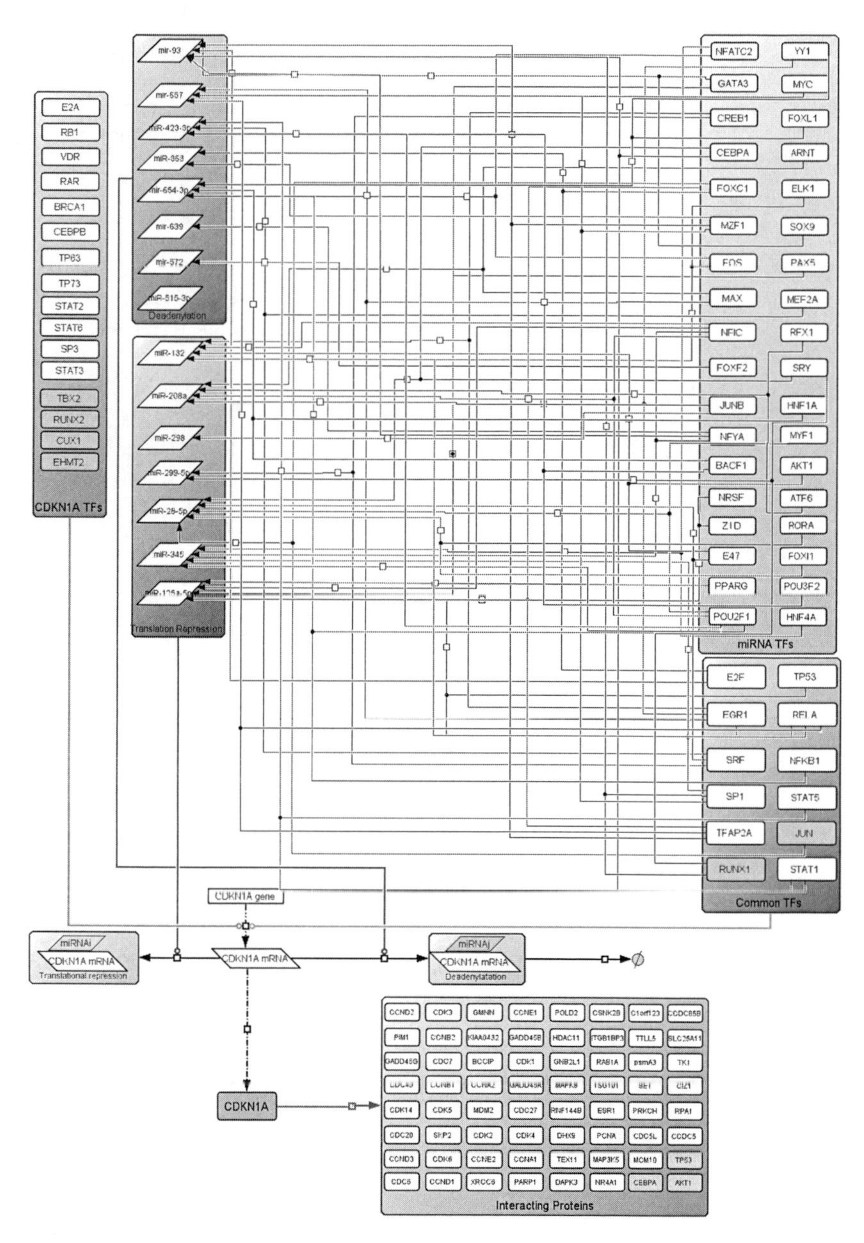

Figure 3.3 Regulatory network of the miRNA target hub *CDKN1A*.
This SBGN compliant illustration shows *CDKN1A* regulating miRNAs that are grouped based on their regulatory mechanism (target deadenylation and target translation repression). Of note, other *CDKN1A* regulating miRNAs exist (Borgdorff *et al.*, 2010; Wu *et al.*, 2010). However, we

considered only those for which consistent quantitative data on their *CDKN1A* repression efficiency was available. The other boxes contain (i) *CDKN1A* TFs, (ii) miRNA regulating TFs, (iii) and those which have *CDKN1A* and a miRNA as target. The latter group is involved in miRNA mediated FFLs. Furthermore, all CDKN1A protein interaction partners are clustered in the box at the bottom. Three proteins are tinted in grey due to their function as TF for *CDKN1A*-regulating miRNAs. These proteins are thus constituents of FBLs. Details on all molecular interactions (CDKN1A-protein, TF-*CDKN1A*, TF-miRNA and miRNA-*CDKN1A*) shown in the network are also listed in Table C.1 in the Appendix along with their corresponding confidence scores.

The network reveals a complex and highly interconnected regulation of the miRNA target hub *CDKN1A* including a large number of regulatory loops. We analysed the network to identify characteristic features. Some results of this analysis are given in Table 3.2. The network has a comparibly low clustering coefficient $C = 0.03$, which is due to the centralized structure of the network around the target hub gene *CDKN1A*. Moreover, there is no interaction among miRNA molecules and we also did not consider any interactions among the TFs. These facts also have direct influence on the characteristic path length ($L = 2.789$), the radius and diameter ($r = 2$; $d = 5$) of the network. For analysing the robustness of the network, we fitted a power law of the form $y = a \cdot x^{-b}$ to the in- and out-degree distribution of the network vertices, where y denotes the number of vertices and x the degree. The results show that, both, the in- and out degree distribution approximate a power law (see Table 3.2) which suggests that the *CDKN1A* regulatory network has the topology of a scale free network, which represents a class of networks that is on average robust against single random structural permutations (Albert *et al.*, 2000; Barabási and Oltvai, 2004).

Network properties		**Network robustness**	
Parameter	**Value**	**Power law of in-degree distribution**	
Number of nodes	144		
Number of edges	198	$y = 7.327 \cdot x^{-0.631}$	$R^2 = 0.364$
Clustering coefficient	0.03		
Network diameter	5	**Power law of out-degree distribution**	
Network radius	2		
Characteristic path length	2.789	$y = 22.327 \cdot x^{-1.018}$	$R^2 = 0.571$
Avg. number of neighbours	2.694		

Table 3.2 Parameters and topological properties of the *CDKN1A* regulatory network.
All these network characteristics were derived using the Cytoscape App NetworkAnalyser (Assenov *et al.*, 2007).

Furthermore, we implemented a network graph in Cytoscape, which illustrates properties of the involved molecules (see Figure C.1 in the Appendix). Thereafter, we analysed the network for motifs and thereby identified a total of 28 regulatory loops (including 25 FFLs and 3 FBLs). All the identified network loops are listed in Table 3.3.

	TFs	Target miRNAs	Biological processes
Coherent FFLs	*RUNX1*	miR-299-5p, miR-93	hemopoiesis
	JUN	miR-28-5p	angiogenesis, cell cycle, proliferation, apoptosis, immune response
Incoherent FFLs	*E2F1*	miR-363, miR-93	apoptosis, proliferation, cell cycle
	TP53	miR-125a-5p	apoptosis, proliferation, cell cycle, DNA damage response, cell senescence
	EGR1	miR-125a-5p, miR-132, miR-208, miR-657	proliferation
	STAT1	miR-93, miR-423-3p	proliferation, apoptosis
	SP1	miR-345, miR-657, miR-93	regulation of transcription
	RELA	miR-132, miR-345, miR-657	Proliferation, apoptosis, immune response, inflammatory response
	SRF	miR-28-5p, miR-299, miR-423	cell senescence, cell migration
	NFkB1	miR-654-3p	apoptosis, immune response, inflammatory response
	TFAP2A	miR-125a-5p, miR-657, miR-93	apoptosis
FBLs	*CEBPA*	miR-28-5p, miR-93	proliferation
	TP53	miR-125a-5p	apoptosis, proliferation, cell cycle, DNA damage response, cell senescence

Table 3.3 Regulatory loops identified in the *CDKN1A* network.
Among the TFs, those experimentally verified to activate *CDKN1A* expression are underlined. MiRNAs with experimental evidence of being regulated by a given TF are also underlined. Predicted TF-miRNA interactions are not underlined. In the right-most column the biological processes are listed in which the TFs are involved according to their GO terms associations. For illustrations of typical miRNA involved regulatory loops and their characteristic time-dependent concentration dynamics see Figure 2.7-Figure 2.10 in Chapter 2.

In the case of the FFLs detected in the *CDKN1A* regulatory network we can differentiate between coherent ($n = 3$) and incoherent FFLs ($n = 22$). In the case of a coherent FFL, *CDKN1A* expression is consistently regulated by a TF which directly inhibits *CDKN1A* expression and in addition represses it indirectly through the activation of a *CDKN1A* repressing miRNA (see Figure 2.9 for an illustration of a coherent FFL). In the case of an incoherent FFL, *CDKN1A* expression is inconsistently regulated by a TF which directly activates *CDKN1A* expression but indirectly represses it through the simultaneous activation of a *CDKN1A* repressing miRNA (Figure 2.10). The complementary scenarios for coherent FFLs (TF → *CDKN1A* ⊢ miRNA ⊢ TF) and incoherent FFLs (TF ⊣ *CDKN1A* ⊢ miRNA ⊢ TF) are also possible, but could not be identified in the *CDKN1A* regulatory network.

Those motifs in which a CDKN1A protein-protein-interaction partner also acts as TF of a *CDKN1A* regulating miRNA were considered as FBLs. All TFs in the *CDKN1A* regulatory network were associated with cellular processes based on their GO terms. Thereby we identified sets of TFs associated with cell proliferation, apoptosis, immune response, inflammation response, cell cycle control, DNA damage response, cell senescence, DNA repair and cell motility and migration (Table 3.3).

The hypothesis of collective target gene regulation (Friedman *et al.*, 2013) is supported by the fact that in our network certain TFs can trigger the expression of several miRNAs in parallel. To know more about the biological processes that the TFs in the regulatory network are involved in and thereby to be able to predict and conclude about the expression of *CDKN1A* in these processes we derived the GO terms (w.r.t. biological processes) associated with each TF using the AmiGO Gene Ontology search tool. GO terms that occur frequently in association with the TFs in our network include cell proliferation, apoptosis and immune response. For some miRNAs we can observe that their TFs are involved in many cell biological processes, e.g. miR-345 (three TFs, six processes), while for others their TFs concentrate on few processes only, e.g. miR-93 (seven TFs, three processes). This suggests that miRNAs whose expression is triggered in many processes play a *housekeeping* role, whereas those belonging to the other group are more likely process-specific miRNAs. As part of our miRNA target hub study, we predicted, by model simulations, the expression of *CDKN1A* in different biological processes. This is described in more detail after the explanation of the model derivation and calibration process.

3.4.2 A predictive model of *CDKN1A* regulation by multiple miRNAs

From our *CDKN1A* regulatory network we derived a kinetic model composed of ODEs based on the system described in Section 3.3.4. The model describes the time evolution of *CDKN1A* mRNA and protein, the miRNAs that regulate *CDKN1A* expression and the duplexes that emerge from hybridizing mRNA and miRNA molecules. Overall, the model consists of 32 time-dependent variables and 64 parameters. For the model parameterization we used the experimental data on target repression efficiencies of *CDKN1A* regulating miRNAs taken from (Wu *et al.*, 2010) and data about miRNA and protein half-lives derived from the literature. Other parameters were assumed or estimated by using data fitting techniques (Balsa-Canto *et al.*, 2010; Vera *et al.*, 2010). The set of ODEs and a detailed description of the model parameterization process are given in Appendix C.2.

Next, we expanded our model to also account for the phenomenon of cooperative target regulation induced by pairs of miRNAs with binding sites that reside in close proximity in the 3′ UTR of a mutual target mRNA (Doench and Sharp, 2004; Saetrom *et al.*, 2007). Therefore, we first predicted the binding sites for miRNAs that are known to regulate the *CDKN1A* expression. Saetrom and colleagues (2007) identified a range (13-35nt) for the seed site distance for which this synergistic effect can be observed. Based on this we identified putatively cooperating miRNA as shown in Figure 3.4. Interestingly, most of the *CDKN1A* regulating miRNAs fulfil the requirements necessary for cooperative target regulation in synergy with at least one other miRNA. The only exception is miRNA-639 because its binding site resides in a rather isolated location. The alignment of each miRNA to the respective *CDKN1A* target site is provided in Table C.2.

Human CDKN1A - 3′ UTR

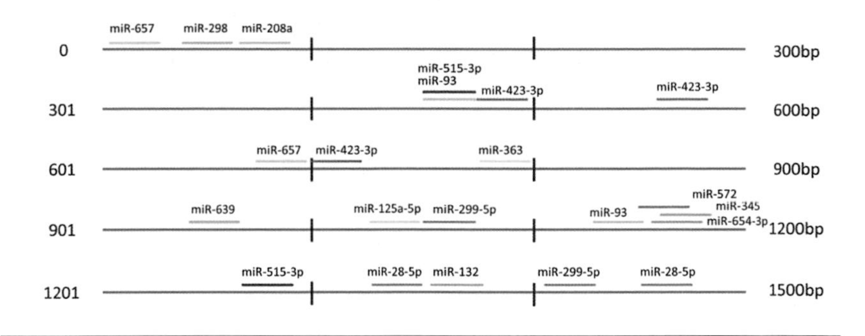

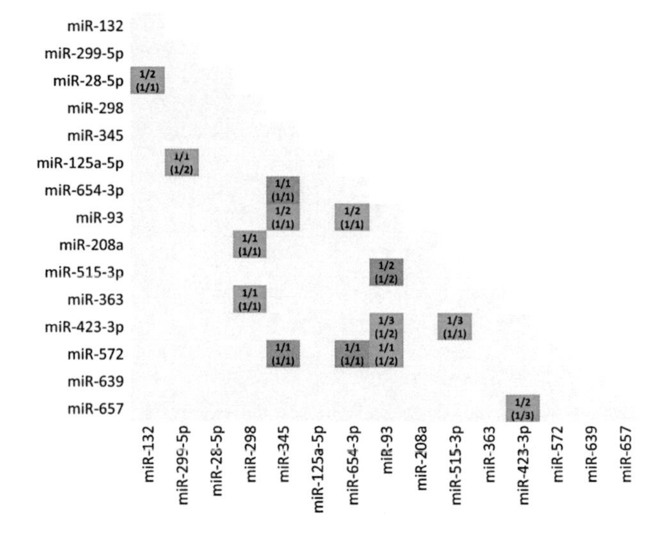

Figure 3.4 Collective and cooperative regulation of *CDKN1A*.
Top: Location of miRNA binding sites in the 3′ UTR of the *CDKN1A* mRNA. Binding sites were predicted using the miRanda target prediction algorithm. **Bottom**: The matrix specifies which miRNA pairs may cooperate based on the seed site distance, which is given in case of a 13-35nt distance according to Saetrom *et al.* (2008). Blue cells at the intersection of a miRNA pair indicate cooperation, while red cells indicate possible competition for mutual binding sites or sites that overlap largely. All other cells suggest an independent (non-cooperative) target repression by both miRNAs. Inside the coloured cells, figures denote the fraction of binding sites interacting with the designated partner for the miRNA on the y-axis, whereas the figure in parentheses is associated with the miRNA on the x-axis.

Next, we revised the equation that describes the time evolution of the *CDKN1A* mRNA and added a term that accounts for the formation of a triplex composed of two cooperating miRNAs (miR_i and miR_j) and the target hub mRNA (th_m). Also we included in our ODE system equations which describe the time-dependent dynamics of all possible triplexes composed of two cooperating miRNAs and the target hub mRNA ($[th_m|miR_i|miR_j]$):

$$\frac{d}{dt} th_m = k_{syn\_th_m} \cdot f_{act}(TF_{th_m}) - th_m \cdot \left(k_{deg\_th_m} + \sum_i k_{ass\_miR_i} \cdot miR_i + \sum_{i,j} k_{comp\_miR_{i,j}} \cdot miR_i \cdot miR_j \right)$$

$$\frac{d}{dt} [th_m|miR_i|miR_j] = k_{comp\_miR_{i,j}} \cdot th_m \cdot miR_i \cdot miR_j - k_{deg\_comp_{i,j}} \cdot [th_m|miR_i|miR_j]$$

For these RNA triplexes we considered two processes: (i) the association of CDKN1A with miR_i and miR_j, represented by the rate constant $k_{comp\_miRi,j}$, and ii) the degradation of the complex, represented by the rate constant $k_{deg\_compi,j}$. MiRNA cooperativity leads to an enhanced target repression, i.e. cooperating miRNA pairs achieve a stronger down-regulation of a mutual target as compared to the collective effects of two individually acting miRNAs. Likewise, the repression efficiency induced by a single miRNA is less even though the concentration is equal to the sum of the concentrations of two cooperating miRNAs. The next chapter describes a more sophisticated computational workflow that we designed to identify pairs of cooperating miRNAs and their mutual targets as well as a case study for the identification of RNA triplexes in the human genome (Schmitz *et al.*, 2014).

3.4.3 Three modes of combinatorial target regulation

Based on the previous paragraph we can assume two modes of combinatorial target regulation by pairs of miRNAs: (i) *synergistic target regulation*, when seed binding sites reside in close proximity, and (ii) *independent target regulation*, when the first condition is not met or in case binding site vicinity has no effect on repression efficiency. In addition, we considered a third possible mode, namely (iii) *interdependent target regulation*. In this mode both miRNAs are dependent on the presence of the other miRNA to achieve an effective target repression. In order to evaluate possible consequences induced by these three modes of concerted target regulation, we adapted the relevant model parameters and simulated for each mode target steady-states for a defined range of miRNA expression values. More specifically, for the mode of *synergistic target regulation* we defined for the target mRNA a complex association rate $k_{comp\_miR_{i,j}} > 0$. The involved miRNAs also had association rates larger than zero ($k_{ass\_miR_i} > 0$; $k_{ass\_miR_j} > 0$). For the *independent target regulation* mode we simply set the complex association rate of the mRNA $k_{comp\_miR_{i,j}} = 0$ and thereby prevented RNA triplex formation *in silico*. Finally, to be able to simulate the third possible mode, i.e. the *interdependent target regulation*, we set $k_{ass\_miR_i} = 0$, $k_{ass\_miR_j} = 0$, and $k_{comp\_miR_{i,j}} > 0$ to ensure that effective target regulation only happens when both miRNAs hybridize with the target mRNA at the same time. See the Appendix C.3 for details on the model parameterization for the purpose of simulating these three scenarios.

Subsequently, we simulated each mode using as example the miRNA pair miR-93 and miR-572. In Figure 3.5, contour plots of CDKN1A steady-states for each mode illustrate how CDKN1A protein levels change for different initial concentrations of the involved miRNAs. In order to observe the effects of underexpression and overexpression of the miRNAs, we defined the range for the initial concentration as $miR_{93}(0)$, $miR_{572}(0) \in [10^{-1}, 10^{2}]$.

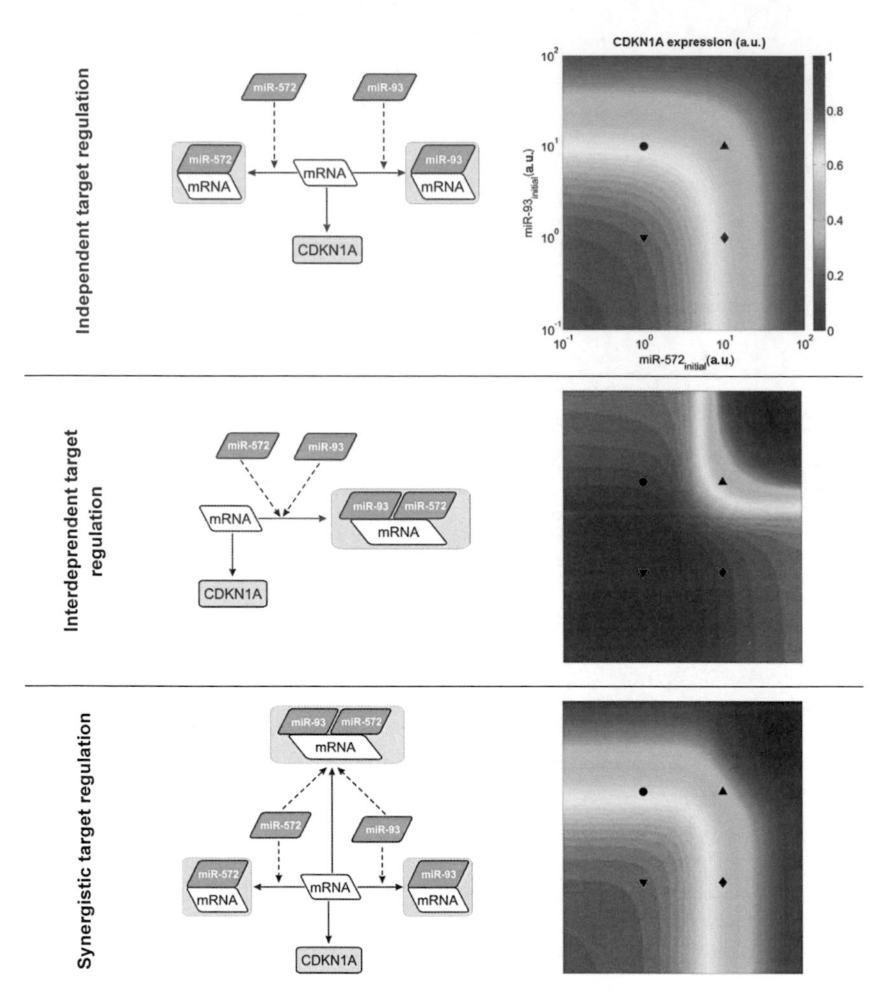

Figure 3.5 Three modes of combinatorial target regulation.
Each column represents one of the three modes of combinatorial target regulation illustrated by a small diagram. Below the diagram simulation results are presented as contour plots of CDKN1A steady-states (a.u.: arbitrary unit) for different initial miRNA concentrations (x- and y-axis). Steady-states were calculated at 48h after miRNA activation as in Wu *et al.* (2010). CDKN1A concentrations are visualized using a rainbow colour scale (see legend) ranging from the colour red (high concentration) to blue (low concentration). As representatives of *CDKN1A* regulating miRNA pairs we selected miR-93 and miR-572. Black symbols within the contour plots are meant as a visual aid. They are positioned at intersection points where both miRNAs have a characteristic value for their initial concentration: (i) both miRNAs are normally expressed ([1, 1], ▼); (ii) miR-572 is overexpressed ([1, 10], ♦); (iii) miR-93 is overexpressed ([10, 1], ●); and (iv) both miRNAs are overexpressed ([10, 10], ▲).

In the independent target regulation mode (Figure 3.5, top), the plot shows that one miRNA can in principle, i.e. when sufficiently overexpressed, silence the target. When both miRNAs are overexpressed, the overall concentration per miRNA required to silence the target reduces slightly. In the case of interdependent target regulation mode (Figure 3.5, middle), the target can only be effectively repressed when both miRNAs are sufficiently overexpressed. This mechanism may explain the relatively poor repression ability of some miRNAs when they are experimentally overexpressed alone (Selbach *et al.*, 2008). Finally, in the synergistic target regulation (Figure 3.5, bottom), effective target repression is achieved when both miRNAs are present in moderate concentrations, whereas if one miRNA is absent it requires a significant overexpression of the other miRNA to achieve an efficient target repression. It can be concluded that the total amount of miRNA required for target silencing drops by approximately one order of magnitude in the case of synergistic target regulation.

3.4.4 Wet lab experiments validate model structure

We tested the validity of our model by performing controlled *in vitro* experiments. To this end, we selected the same pair of putatively cooperating miRNAs (miR-572 and miR-93) that we used for the simulations described in the previous paragraph. Both were transfected individually at a concentration of 100 nM or in combination (50 nM each) to Sk-Mel-147 cells. Furthermore, to ensure sufficient expression of *CDKN1A*, the cells were treated with the anti-cancer drug doxorubicin (250 nM), a genotoxic stress-inducing agent. *CDKN1A* is in this context activated by the transcription factor TP53 as part of the cellular stress response pathway (Riley *et al.*, 2008). After doxorubicin treatment, the expression of TP53 was measured using immunoblotting. This data was used to characterize the transcriptional function of *CDKN1A* using the linear interpolation function in MATLAB. Additionally, CDKN1A protein expression levels in response to stress were measured by immunoblotting at different time points. Thereby, CDKN1A dynamics were obtained for (i) endogenous miRNA expression, (ii) individual miRNA-572 or miR-93 overexpression, and (iii) moderate overexpression of both putatively cooperating miRNAs. Interestingly, the experimental results are in well accordance with our corresponding model simulations as can be seen in Figure 3.6. More details on the experimental procedure are provided in Appendix C.3.

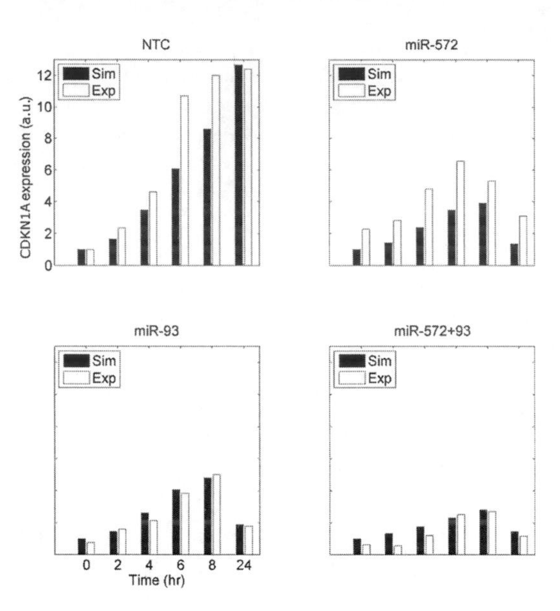

Figure 3.6 Regulation of CDKN1A by miR-572 and miR-93.
In the depicted plots we compare our model simulations (black bars) of CDKN1 repression by miR-572 and miR-93 with the results of the immunoblotting validation experiments (white bars; a.u.: arbitrary unit). The labels of each plot indicate the molecule transfected to the SK-Mel-147 cells, namely a non-targeting control (NTC), miR-572, miR-93, and both miRNAs in combination. Prior to transfection, cells were treated with doxorubicin which triggers *CDKN1A* expression via TP53 due to the induction of genotoxic stress.

The model simulations as well as the experimental results illustrate well how CDKN1A expression is increasing due to the induced stress signal that triggers *TP53* expression. However, the effect of the transfected miRNAs overcomes the stress-induced CDKN1A up-regulation and becomes visible 24 hr after treatment. In the case of miR-93 transfection, we witnessed stronger target repression efficiency as compared to miR-572. However, in the case of combined miR-572 and miR-93 transfection (each at half concentration) one can observe a dampened increase in CDKN1A concentration at the early time points, followed by a twofold drop after 24 hr. These results support our model structure and the predicted cooperativity between miR-572 and miR-93.

3.4.5 Model simulations predict processes-specific CDKN1A expression

After calibrating the kinetic model and after validating its predictive capacity and structure validity we used the model to predict the expression of CDKN1A in the context of different cell biological processes. More specifically, based on the biological processes associated with the TFs of the *CDKN1A* regulating miRNAs we performed simulations for these processes and compared the predicted CDKN1A expression levels with literature derived qualitative measures. Toward this, we first extracted the GO terms associated with the TFs of the miRNAs, assuming that a miRNA is expressed in a given biological process when at least one of its trans-acting regulators is associated with this process. Subsequently, we predicted the CDKN1A steady-state levels for the investigated biological processes (DNA repair, cell migration, DNA damage, senescence, inflammatory response, immune response, cell cycle, apoptosis, and cell proliferation) with and without considering the phenomenon of cooperatively acting miRNAs. Figure 3.7 illustrates in a matrix (left) the process-specific activity profiles of *CDKN1A* regulating miRNAs as well as the simulation results (right) which represent predicted steady-states of CDKN1A protein in the respective processes. Our predictions are supported by recent experimental evidence. For example, transcriptional activation of several miRNAs can relieve cells from CDKN1A-mediated senescence (Borgdorff *et al.*, 2010). In line with this, our model predicted high CDKN1A levels during senescence and low levels in cell proliferation, a process in which many *CDKN1A*-targeting miRNAs are activated (Figure 3.7).

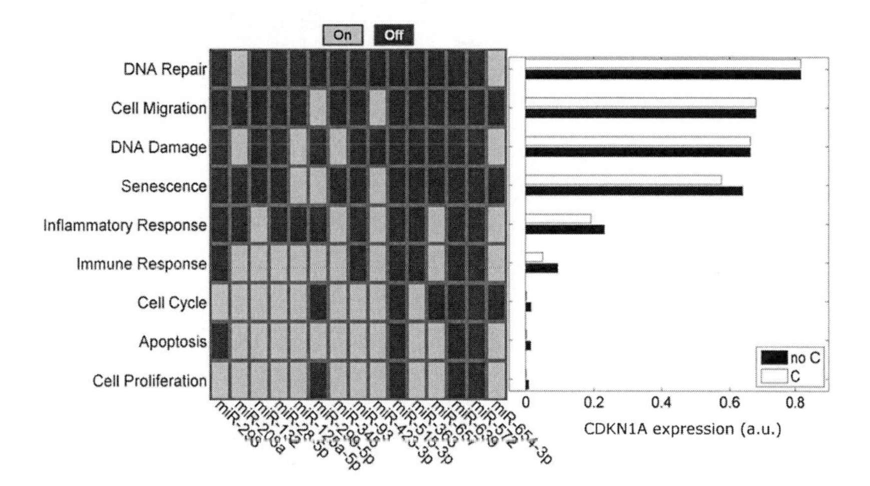

Figure 3.7 MiRNA activity profiles and CDKN1A expression in different biological processes. Left: The states (on/off) indicating miRNAs activation for different biological processes are inferred from the GO terms of their TFs. **Right:** Predicted process-specific CDKN1A steady-state levels considering and not considering synergistic target regulation by miRNAs, represented by white bars and black bars respectively (a.u.: arbitrary unit). The rows are ordered based on the predicted CDKN1A expression levels not considering the contextual closeness of the biological processes.

For the apoptosis process our model predicts a low CDKN1A expression level. One possible event triggering apoptosis initiation is the mitotic catastrophe in which the cell-cycle fails to be arrested by the cell (Vitale *et al.*, 2011). Mitotic catastrophe is initiated when mediators of cell cycle arrest, e.g. CDKN1A and 14-3-3-σ, are insufficiently expressed (Chan *et al.*, 2000; Schultz *et al.*, 2009). Many of our predicted miRNA transcription factors ($n = 15$) are active in apoptosis and therefore a large number of

CDKN1A regulating miRNAs is active (11 out of 15) causing a strong *CDKN1A* repression in apoptosis which may facilitate mitotic catastrophe. More details and all parameter values used in simulations of processes-specific CDKN1A expression are provided in Appendix C.3 and Table C.5. Of note, cooperative miRNA regulation affects only some of the biological processes, namely senescence, inflammatory response, and immune response, while in other processes the synergistic effect of miRNA regulation is either negligible or not present. In these contexts either the expressed miRNAs do not cooperate because their binding sites are not in close proximity or CDKN1A is (almost) silenced and therefore miRNA cooperation makes no difference. Overall, it can be concluded that a fine-tuned target regulation achieved through distinctively and cooperatively acting miRNAs may be selectively induced for different cell biological scenarios.

3.4.6 Prediction of *CDKN1A* expression for different tissues

After predicting *CDKN1A* expression in different cell biological processes we tested if our model is also able to predict tissue-specific *CDKN1A* expression. To this end, we extracted from the miRNAMap database (release 2.0; Hsu *et al.*, 2008) miRNA expression profiles from quantitative PCR experiments for twelve different tissues and normalized them according to the most abundant miRNA observed in each tissue. We used these expression levels to define the initial miRNA concentrations in our model prior to simulation. Thereafter, we predicted *CDKN1A* steady state levels for twelve human tissues and made a qualitative comparison with tissue-specific *CDKN1A* expression levels derived from the ArrayExpress database (version as of January 2012; (Brazma, 2003). To make the results comparable with the publicly accessible *CDKN1A* tissue profiles we discretized the simulation results to finally differentiate between two ordinal categories (low expression and high expression). This was done to account for the variations in the experimental data that depend on quality, platform and normalization methods used. More details on the discretization step are provided in Appendix C.3. In 9 out of 12 cases the model simulations correctly predicted the tissue-specific *CDKN1A* expression. This is a rather good result considering the fact that other factors can influence the experimental outcome. Controlled validation experiments in purified cell cultures would be required to come up with a more conclusive estimation of the predictive capabilities of our *CDKN1A* regulation model.

3.4.7 Modulation of miRNA cooperativity influences target repression efficiency

Our experiments have demonstrated that miR-93 and miRNA-572 synergistically regulate the expression of *CDKN1A*. Although we assume that other pairs of miRNAs can also cooperate (see next chapter) we continued our analysis with this specific pair for which we derived a well-characterized and parameterized kinetic model. We were then interested whether the strength of cooperativity, which we named as K, may have an influence on the target repression efficiency. To this end, we defined an interval $K = k_{comp\_miR_{i,j}} \cdot [10^{-5}, 10^{5}]$ and simulated the model applying K in steps of $10^{0.25}$. For each step we determined the CDKN1A protein expression level at 48 hr (Figure 3.8, left). The simulations were performed for four different scenarios representing different initial miRNA concentrations: (i) both miRNAs are normally expressed, (ii) and (iii) one of the two miRNAs is overexpressed, while the other miRNA is normally expressed, and (iv) both miRNAs are overexpressed. In Figure 3.8 (left) each scenario is depicted using a different symbol (see also the figure legend).

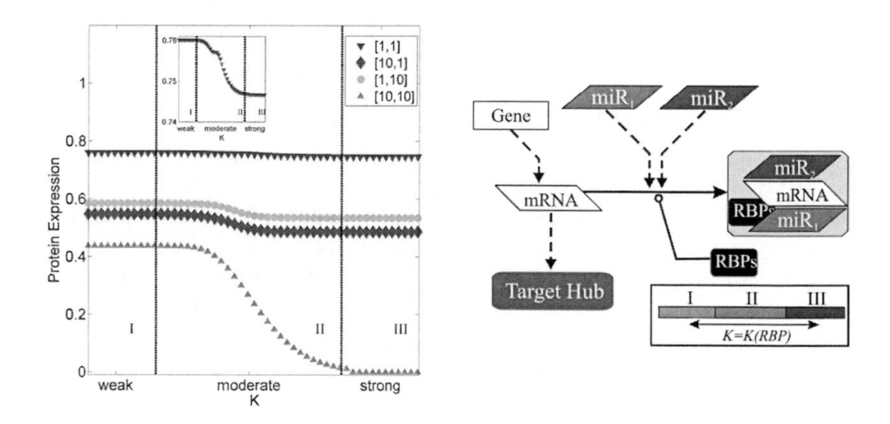

Figure 3.8 Target repression efficiency depending on the strength of miRNA cooperativity (*K*). **Left**: Data points illustrate CDKN1A protein expression levels at 48 hr of simulation for increasing cooperation strength (x-axis). The symbols used represent different initial miRNA concentrations: (i) both miRNAs are normally expressed ([1, 1], ▼); (ii+iii) one miRNA is overexpressed ([1, 10], ♦ or [10, 1], ●); and (iv) both miRNAs are overexpressed ([10, 10], ▲). The small internal plot represents a zoom into a more narrow range of protein expression (y-axis, [0.74,0.76]) to illustrate that also the data points represented by (▼) follow a sigmoid shaped curve. **Right**: Illustration of the hypothesis that RBPs can fine-tune the repression efficiency of cooperating miRNAs ($K = K(RBP)$).

For visual aid we divided the simulation results into three zones representing different strength of the cooperativity K. In zone I, the synergistic effect of cooperative target regulation is negligible, i.e. the target repression induced by the pair of miRNAs is comparable to that achieved in the independent target regulation mode. In zone II, increasing values of K induce a visible effect on target repression efficiency, especially in the scenario where both miRNAs are overexpressed (purple triangles). Finally, in zone III the target can be completely silenced when both miRNAs are overexpressed (purple triangles), whereas the other scenarios do not differ compared to zone II.

Of note, the target repression curves for one overexpressed miRNA (green and red) differ because of their individual target repression efficiencies ($k_{ass_miR_{93}}$and $k_{ass\_miR_{572}}$). In summary, we found that active modulation of miRNA cooperativity can heavily influence the efficiency of target repression. The highest sensitivity to changes in the cooperativity strength can be observed in scenarios where both miRNAs are overexpressed. Possible factors that may influence target repression efficiency are for example some RNA binding proteins (RBPs) like FMRP or PUF which are known to interact with the RNA induced silencing complexes (RISC, Figure 3.8). In contrast, other RBPs (e.g., DND1 and HuR) were reported to counteract miRNA-mediated repression (Krol *et al.*, 2010).

This and the results from our model simulations suggest that target repression can be fine-tuned by mechanisms of combinatorial and cooperative miRNA regulation as well as the influence of RBPs. Furthermore, the effect of synergistic target regulation by pairs of miRNAs on a mutual target mRNA may be tuned by RBPs which can shift the effective target repression performance from one zone to the other (Figure 3.8, right).

3.4.8 MiRNA cooperativity to enhance noise buffering

In the last step of our investigation of combinatorial target regulation by several partially cooperatively acting miRNAs, we studied whether changes in the miRNA abundance and cooperativity can modulate the response of a target hub gene to transient stimulation, such as in the case of TP53-induced transcriptional activation of CDKN1A in response to DNA damage. We performed a series of simulations, in which we altered μ and τ, representing the amplitude and time duration of a transient signal, and computed the peak of the CDKN1A response (Figure C.5A, right). The simulation experiments are described in detail in Appendix C.3. We observed that for short-term stimulation ($\tau = 1\,hr$), the CDKN1A response peak correlates with μ, the signal amplitude, whereas in case of long-term stimulation ($\tau = 10\,hr\ or\ 24\,hr$) and high signal amplitudes ($\mu > 1$), CDKN1A responses converge with those that would emerge when no miRNA regulation was involved (Figure C.5B, middle and right). This behaviour can be explained by the accelerated consumption of free miRNAs due to the continuous synthesis of *CDKN1A* mRNA molecules. The simulations further show that, in general, higher signal amplitude is required to achieve a significant CDKN1A response under miRNA regulation. This effect is even enhanced through miRNA cooperativity. However, prolonged signal duration can counterbalance this effect. We conclude that higher abundance of miRNAs as well as miRNA cooperativity can enhance buffering of signalling noise (Herranz and Cohen, 2010).

3.5 Summary

It is established knowledge that miRNAs are important post-transcriptional regulators of gene expression and that they are involved in many cellular processes, e.g. cell cycle, proliferation, apoptosis, migration, and angiogenesis, and in the emergence and progression of human diseases like cancer (Schmitz *et al.*, 2013). The latter is often mediated or even caused by the dysregulation of miRNA expression. More than 2,000 human miRNAs have been identified so far (Kozomara and Griffiths-Jones, 2014) and it has been shown that each miRNA has dozens to hundreds of target mRNAs due to the incomplete sequence complementarity required for target recognition and their short length (Selbach *et al.*, 2008).

In this study we have addressed the to date neglected phenomenon of combinatorial target regulation by many miRNAs sharing a mutual target gene. One exception is the study presented in (Wu *et al.*, 2010) where authors investigated the regulation of the miRNA target hub *CDKN1A* (in that study referred to as p21). We used this as a starting point for our own in-depth analysis, in which we analysed consequences of combinatorial miRNA regulation on target hub gene expression. Furthermore, we investigated the phenomenon of cooperative target regulation by pairs of miRNAs, which was previously described by Doench and Sharp (2004) and Saetrom *et al.* (2007).

The outlined analysis pipeline, designed and implemented to study the combinatorial regulation of the target hub *CDKN1A* by miRNAs, can be used as a template for studying other miRNA target hubs or other miRNA-involving regulatory networks. The validity of the network and the kinetic model was proven by the described wet lab experiments in which we validated cooperative *CDKN1A* regulation by miRNA-93 and miR-572. Furthermore, for the predictions of CDKN1A expression in different cell biological processes we found supportive evidence in the literature.

Our model of *CDKN1A* target hub regulation can be utilized to predict *CDKN1A* expression changes and analyse consequences, e.g. in the context of cancer or other diseases. For example, *CDKN1A* has two possible roles in the cell cycle: (i) under normal conditions it is expressed at low levels because it inhibits cell cycle progression; (ii) in response to stress CDKN1A is up-regulated through TP53-dependent pathways and functions as a cell cycle inhibitor (Jung *et al.*, 2010). According to our network, TP53 can activate *CDKN1A* expression directly but also inhibits *CDKN1A* expression indirectly via the activation of a *CDKN1A* suppressing miRNA (miR-125-5p; see Table 3.3); this is a working example of an incoherent FFL. Interestingly, this FFL has important consequences for *CDKN1A* regulation in cancer conditions. In non-small lung cancer, for example, miR-125-5p is down-regulated which accounts for a deactivation of the FFL (Jiang *et al.*, 2010b). Our simulations suggest that this condition can favour cancer progression and chemo-resistance by delaying the initiation of CDKN1A-triggered cell cycle arrest after DNA damage (Figure C.4).

We also considered the recently discovered phenomenon of cooperative target regulation by pairs of miRNAs in our investigation of target hub regulation. Experimental evidence for this sophisticated mechanism of post-transcriptional gene regulation was produced in (Doench and Sharp, 2004) and in (Saetrom *et al.*, 2007). They have shown that miRNAs with binding sites which reside in close proximity can cooperate and induce an enhanced target repression capacity. Using computer simulations we have shown that target repression efficiency can be controlled by modulating the strength of cooperativity between two miRNAs (Figure 3.8) and that this strength is likely to be tuned by RBPs. However, there is still little knowledge about this phenomenon, especially in terms of other possibly cooperating miRNA pairs. Therefore, we developed a workflow for the prediction and analysis of cooperating miRNA pairs and their mutual target genes which is discussed in detail in Chapter 4.

3.5.1 Confidence scores for molecular interactions

In the first step of the CATH workflow, for the reconstruction of a miRNA target hub network we integrated heterogeneous data from multiple resources focusing on molecular interaction data. These data originate from experiments on various platforms, computational predictions and the literature. Although data integration is an essential step for the analysis of biological processes in general, and in the analysis of miRNA target hubs in particular, the inconsistency in the way the data were generated may lead to errors or false assumptions. Quality control is therefore an indispensable step in data integration (Santra *et al.*, 2014). To assess the reliability of molecular interaction data it is advisable to assign a confidence score to each interaction.

To date no gold standard for this purpose exists, however several databases of molecular interaction data have implemented their own customized scoring systems. One example is the ConsensusPathDB database in which the confidence score is computed based on GO terms, pathway annotations and network topological features (Kamburov *et al.*, 2011). In the SignaLink2 database, PPIs are scored based on the GO semantic similarity of the interaction partners (Fazekas *et al.*, 2013).

Despite database specific scoring systems, some generic scoring systems have been proposed, for example in (Deane *et al.*, 2002), where the authors developed a scoring system for protein interactions which relies on similarities in mRNA expression profiles. Bader *et al.* (2004) also used expression correlation coefficients but added functional similarity as a criterion in their scoring system. Li and co-authors proposed a confidence scoring system which they refer to as PRINCESS (protein interaction confidence evaluation system with multiple data sources), which is based on a Bayesian network approach and integrating evidence from model organisms, functional annotations, gene expression data, genomic context, and network topology (Li *et al.*, 2008). Like this system, all other previously proposed scoring schemes are designed for protein-protein interactions only, some of them even relying on a specific sort of data. Therefore, these scoring systems cannot be easily adopted for heterogeneous interaction networks established from different sources. For this reason, we decided to implement our own scoring system for molecular interactions in a miRNA target hub network, which can also be adopted for other networks, like signalling pathways, PPI networks or gene regulatory networks. We found inspiration in the MIscore system employed for scoring PPIs in the IntAct database (Kerrien *et al.*, 2012), which is based on common and minimum curated information reporting a molecular interaction experiment. It is composed of weighted scores for the experimental detection method, the interaction type and the number of publications in which evidence for this interaction was reported. Details on how we adopted this system for the miRNA target hub network were explained in Section 3.3 on the Research strategy.

3.5.2 Mathematical models of miRNA-target regulation

Models describing the general mechanism of miRNA-target regulation have already been proposed. The first model on miRNA-mediated gene silencing was published by Levine and colleagues (Levine *et al.*, 2007) and expanded by Nissan and Parker (2008) and Zinovyev *et al.* (2010). Each model added more details about the mechanisms involved in miRNA-induced target repression, but miRNA cooperativity was still not examined. We developed the first kinetic model for the phenomenon of miRNA target hub regulation involving miRNA cooperativity. In our model we used a simplified description of miRNA-mediated target repression but considered, in contrast to the other models, collective regulation of a mutual target by multiple and cooperative miRNAs. We used quantitative data derived from PCR experiments and Western blots from (Wu *et al.*, 2010) to calibrate our model and to predict quantitative effects on target mRNA and protein levels. This is an improvement on the other data-driven models of miRNA-regulation as, for example, the one developed by Khanin and Vinciotti (2008). Their model on miR-124a induced post-transcriptional gene regulation was calibrated using semi-quantitative data from microarray data and describes subsequent changes on target mRNA levels only. In contrast, our model incorporates multiple levels of regulation, i.e. transcriptional and post-transcriptional *CDKN1A* regulation, and transcriptional regulation of *CDKN1A* targeting miRNAs.

In another more recent study, we used the same network modelling approach to analyse the gene regulatory network involving miR-205 with respect to targets that are important in the response to anti-cancer drugs and which may play a role anti-apoptotic processes. In this context, we simulated the interplay between miR-205 and E2F1 plus downstream targets to investigate the role of miR-205 and E2F1 in cancer progression, aggressiveness and, most importantly, in chemo-resistance (Vera *et al.*, 2013). Similarly, we analysed the role of miR-638 in melanoma metastasis (Bhattacharya *et al.*, 2015). In this model, we describe the dynamics of the network components and their non-linear characteristics caused by the double-negative feedback loop between miR-638 and TPAP2A. This is in line with studies such as the one from Lee and colleagues in which they constructed a network-model of the tumour suppressor miRNA miR-204 and 18 target genes important for tumour progression (Lee *et al.*, 2010). In the present model we did not analyse consequences on disease emergence and progression but successfully predicted CDKN1A expression for nine different cell biological processes and twelve different human tissues.

3.5.3 Mechanisms of miRNA-target regulation

So far, three general mechanisms of miRNA-target regulation have been described: (i) *translation repression* by blocking of translation initiation, elongation blocking or early translation termination (e.g., through ribosome drop off); (ii) *deadenylation* and subsequent target degradation; and in rare cases (iii) RISC-induced target *cleavage*, a phenomenon more frequently found in plant RNA-induced gene silencing (Filipowicz *et al.*, 2008; Jones-Rhoades *et al.*, 2006; Figure 1.4). Levine *et al.* (2007) suggested in addition, that miRNA-specific parameters may determine the effect of miRNA-regulation on target mRNA levels, but also global parameters (controlling mRNA turnover) and circumstances in cell biological processes in which target repression occurs. Here we investigated collective action of many miRNAs on a mutual target gene and synergistic regulation induced by pairs of miRNAs which at the same time supports and extends

Levine's notion. These different *modi operandi* facilitate a fine-tuned control of target gene expression as well as cell type and process-specific target regulation. The latter was investigated in our study by integrating knowledge on the role of miRNA and *CDKN1A* regulating TFs in different biological processes (determined by GO analysis). The *CDKN1A* regulatory network constructed by integrating data from various resources (Figure 3.3) indicates that distinctive regulation enforced by different TFs may be one of the mechanisms by which the cell biological processes determine the efficiency of miRNA-mediated repression. In Figure 3.9, two examples of process-specific subnetworks are illustrated, showing TF-induced expression control of *CDKN1A* and its miRNA regulators.

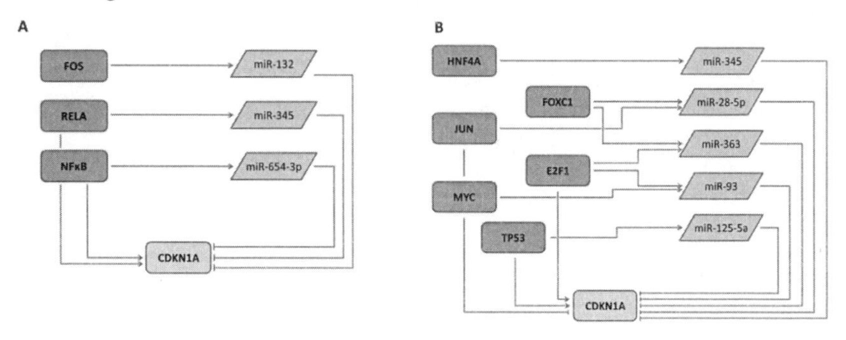

Figure 3.9 Examples of process-specific regulation of *CDKN1A*.
The TFs (dark grey rectangles) regulate *CDKN1A* directly, and indirectly through miRNAs (parallelograms). Illustrated are two process-specific subnetworks: (**A**) inflammatory response and (**B**) cell cycle.

With respect to the phenomenon of synergistic target regulation by pairs of cooperating miRNAs, we confirmed the results from Doench and Sharp (2004) in which miRNA binding sites that reside in close proximity in a targets 3′ UTR facilitate an enhanced target repression. The optimal seed site distance, according to Saetrom *et al.* (2007), is 13-35nt. We accordingly computationally predicted all miRNA pairs possibly cooperating in the regulation of *CDKN1A* (Figure 3.4, top). Our predictions exhibit occurrences of larger motifs including three and more non-overlapping miRNA binding sites that reside in close proximity (e.g., miR-657, miR-298 and miR-208a, see Figure 3.4, bottom) but also clusters of partially overlapping binding sites (e.g., miR-654-3p, miR-572, miR-345 and miR-93). Clustered miRNA binding sites have been previously described by Rigoutsos (2006) and suggest other, more complex, mechanisms inducing non-linear target repression pattern. The often observed mild repression capacity of miRNAs (Selbach *et al.*, 2008) may occur due to the mechanism of interdependent target regulation that is illustrated in Figure 3.5. In this case, the overexpression of a single miRNA cannot facilitate full silencing of the designated target gene, *in vivo* and *in vitro* evidence of which can be found in the article published by Bartel (2004). In contrast, the mechanism of synergistic target regulation can induce sophisticated target repression, where the strength of cooperativity between a miRNA pair may be determined by the interaction of RISC with other RBPs (Vohradsky *et al.*, 2010).

In the analysis of the *CDKN1A* regulatory network we detected recurring network motifs including coherent and incoherent FFLs but also FBLs. In total, 30 regulatory loops were discovered, most of them being FFLs (Table 3.3). Characteristics of coherent and incoherent FFLs involving TFs and miRNAs have been described in several studies (Shalgi *et al.*, 2007; Tsang *et al.*, 2007; Tu *et al.*, 2009; Aguda, 2013). Using these regulatory motifs in conjunction with the phenomenon of synergistic target regulation, fine-tuned transient or long-term process-specific regulatory responses may be facilitated as illustrated in Figure 3.10. This figure demonstrates how process-specific TF activity can modulate duration and intensity of transient peaks in target protein concentration in response to a step-like TF activation (Mangan and Alon, 2003) and in cooperation with a second miRNA (Figure 3.10, bottom right).

We have recently shown that the tumour suppressor TFAP2A is engaged in a double negative feedback loop with the oncomiR miR-638 (Bhattacharya *et al.*, 2015). However, in the here presented study we predicted that TFAP2A forms incoherent FFLs with miR-125a-5p, miR-657 and miR-93 (see Table 3.3) which leads to a nonlinear regulation of their mutual target *CDKN1A*. Of note, in Bhattacharya *et al.* (2015) we found that an epigenetic modification of the miR-638 promoter, which happens to be methylated, interfered with the inhibition of miR-638 by TFAP2A. Therefore, although our proposed *CDKN1A* regulatory network is already quite complex and describes *CDKN1A* regulation in great detail, it does not yet include a layer of epigenetic gene regulation which may further increase its complexity.

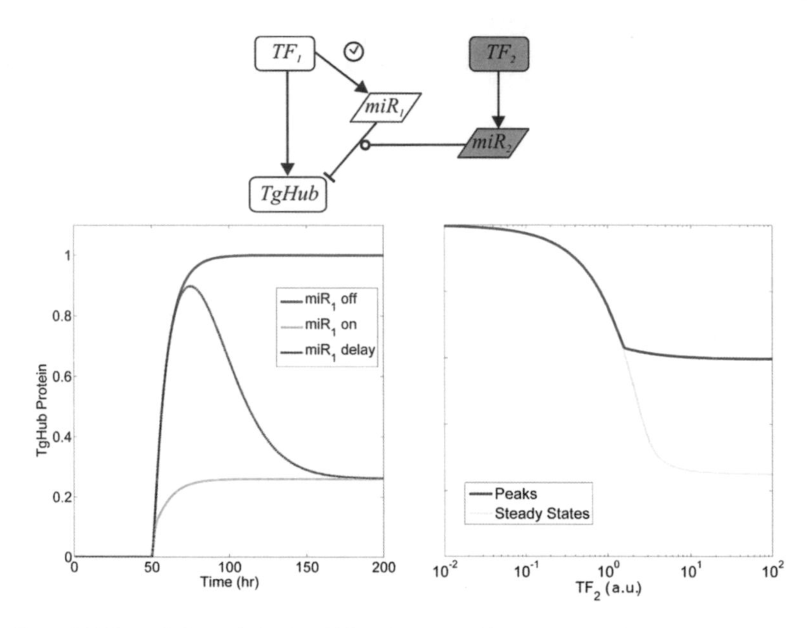

Figure 3.10 Target hub regulation in a FFL without and with cooperating miRNA.
The *CDKN1A* regulatory network exhibits many regulatory loops (mostly FFLs). These loops and the phenomenon of cooperative target regulation by pairs of miRNAs can induce a fine-tuned regulation of target protein expression. In the illustrated example, an incoherent FFL in which the miRNA (miR_1) is expressed with delay, the target hub (TgHub) and the miRNA share the same TF ($TF_1 \rightarrow$ TgHub, $TF_1 \rightarrow miR_1$, $miR_1 \dashv$ TgHub). In the second scenario a second miRNA (miR_2) is activated by an independent TF. Both miRNAs cooperate in the regulation of their mutual target ([miR_1 AND miR_2] $\dashv$ TgHub). The simulations show that, in the first scenario, a step-like activation of TF_1 induces a transient peak in the target hub expression (bottom left, red curve), whereas when $miRNA_1$ is expressed without delay, it induces stable down-regulation of the target hub (bottom left, green curve). On the other hand, the intensity of this peak can be modulated by the activation of TF_2. The subsequently activated $miRNA_2$ mediates the depicted peak and steady state changes (bottom right; a.u.: arbitrary unit).

3.5.4 Generalization of the CATH workflow

The classical approach for the investigation of miRNA-induced target gene repression is to use algorithms for the prediction of miRNA-target interactions. The here proposed workflow extends the idea of solely identifying possible miRNA-target interactions. We suggest building a comprehensive network of target gene regulation involving other miRNAs, TFs and protein-protein interactions in order to derive sophisticated and predictive models (Nikolov *et al.*, 2010; Lai *et al.*, 2012). Shalgi and colleagues (2007) predicted a total of 834 human miRNA target hub genes, each putatively targeted by 15 or more miRNAs. Therefore, our proposed analysis pipeline may be used as a template for the investigation of other important miRNA target hubs.

The established network of *CDKN1A* regulation can be considered as a basis for further experimental investigations. In the derived mathematical model, the equations are characterized with biological knowledge and parameterized with quantitative data. For miRNA-target interactions, this data ideally comes from perturbation experiments, in which expression levels of the critical miRNAs are tuned and consequences for target mRNA and protein concentration are quantified over time. Once an initial version of the model is established, it requires iterative cycles of modelling and experimentation to calibrate it and make it a predictive model. In case data is missing, analytical tools can help to estimate missing parameter values (Alves and Savageau, 2000; Savageau *et al.*, 2009). In the ODE system of *CDKN1A* regulation, we used global and local optimization algorithms to predict these values.

The focus in the next chapter is on the development of a sophisticated computational workflow for the reliable prediction of cooperating miRNAs pairs and their mutual target genes.

4 Prediction and validation of cooperating miRNAs and their mutual targets

This chapter is based on the following publications:

Lai X*, **Schmitz U***, Gupta S, Bhattacharya A, Kunz M, Wolkenhauer O, Vera J (2012) Computational analysis of target hub gene repression regulated by multiple and cooperative miRNAs. Nucleic Acids Res 40: 8818-8834.

Schmitz U, Wolkenhauer O (2013). Web resources for microRNA research. Adv Exp Med Biol, 774, 225–250.

Vera J, **Schmitz U** (2013) Computational MicroRNA Biology. In Encyclopedia of Systems Biology (ed. Dubitzky W, Wolkenhauer O, Cho K, Yokota H), pp. 473-480. Springer New York.

Schmitz U, Lai X, Winter F, Wolkenhauer O, Vera J, Gupta S (2014). Cooperative gene regulation by microRNA pairs and their identification using a computational workflow. Nucleic Acids Res 42 (12): 7539-7552.

(The '*' symbol denotes equal contributions to the respective work.)

Synopsis

In the previous chapter I described an integrative approach for the analysis of miRNA target hub regulation. It was shown that target hub genes can be collectively regulated by many miRNAs, some of which can act synergistically to induce more specific and more efficient target repression. In this chapter I describe an integrative workflow for the ***Pre****diction and validation of* ***Co****operating* ***M****iRNAs and their mutual targets (****PreCoM****). The workflow integrates a wide range of methods from the fields of bioinformatics, structural biology and systems biology. In a case study we predicted triplexes of putatively cooperating miRNAs and their targets in a whole human genome analysis. Each triplex was analysed with respect to structure and thermodynamic stability. Furthermore, a kinetic model of cooperative target regulation was constructed and parameterized individually for each triplex (n>600.000). Thereafter, predictive simulations were performed for determining the strength of cooperativity and target repression efficiency. All human RNA triplexes predicted in this work together with information on their specific characteristics and model simulations are stored in a MySQL database which can be accessed at www.sbi.uni-rostock.de/triplexrna/.*

4.1 Background

As shown in the previous chapter, the concerted target gene regulation by miRNAs realizes sophisticated control functions, e.g. noise buffering or homeostasis, and can facilitate fine-tuned regulation of target expression suitable for the demand in different cell biological processes (Lai *et al.*, 2012; Friedman *et al.*, 2013). Chapter 3 also described how to computationally assess concerted target regulation as well as the phenomenon of cooperative target regulation by pairs of synergistically acting miRNAs. We and others have experimentally proven that cooperating miRNAs can induce a more efficient target repression (Vella *et al.*, 2004; Saetrom *et al.*, 2007; (Lai *et al.*, 2012; Figure 4.1). However, only two incidences of cooperating endogenous miRNAs have been validated so far: (i) the human miRNAs hsa-miR-93 and hsa-miR-572 cooperatively repress the expression of the cell cycle regulator *CDKN1A* (Lai *et al.*, 2012) and (ii) in *C.elegans* pairs of the miRNA cel-let-7 can realize an enhanced repression of lin-41 due to two adjacent binding sites (Vella *et al.*, 2004).

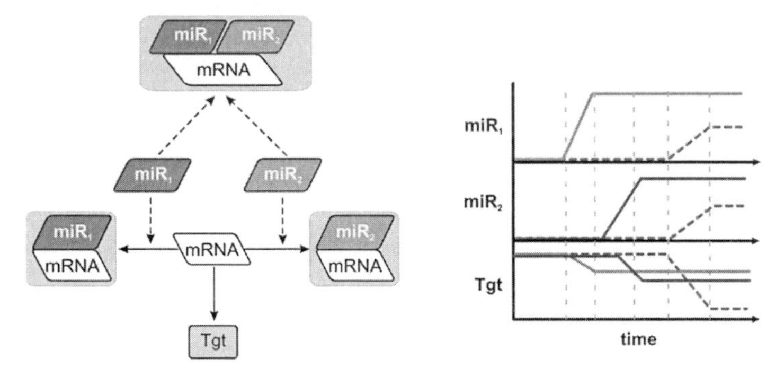

Figure 4.1 Cooperative target regulation through a miRNA pair.
Left: The diagram illustrates possible interactions of the cooperating miRNAs with the target mRNA: (i) one of the miRNAs forms a duplex with the mRNA, (ii) both miRNAs hybridize with the target and form a triplex structure. **Right:** The plot shows the repressive effect induced by single miRNA regulation (solid lines) and by cooperative regulation (dashed lines). Even when the miRNAs are expressed at half concentration, their cooperativity induces an enhanced target repression.

In the year 2007, Hon and Zhang found that the distance between two seed binding sites of a miRNA pair may affect the level of target repression (Hon and Zhang, 2007). In the same year, Saetrom and colleagues specified the range of 13-35nt as optimal distance for cooperative target regulation (Saetrom *et al.*, 2007). This chapter presents a computational workflow (PreCoM) for the prediction of cooperating miRNA pairs and their mutual targets as well as the characterization of the triplex structures formed by these three RNA molecules. Moreover, the workflow also predicts the target repression efficiency. To achieve this goal it requires the implementation of six interdependent computational steps: (I) prediction of miRNA target sites in the 3′ UTR of mRNA targets; (II) identification of neighbouring target sites with a distance between 13-35nt;

(III) secondary structure prediction and analysis of putative RNA triplexes derived from steps I+II; (IV) molecular dynamics simulations (MDS) of RNA triplexes, plus derivation of a thermodynamic profile; (V) prediction of equilibrium concentrations of the inherent RNA monomers and complexes (duplexes and triplexes); and (VI) prediction of target repression efficiency based on kinetic model simulations (Figure 4.2).

In a case study, we implemented the PreCoM workflow and predicted for the entire human genome putative targets of cooperative miRNA regulation. The results of our analysis suggest that thousands of genes can be synergistically regulated by miRNA pairs. We designed a database named TriplexRNA to make these results available, which, apart from the predicted triplexes, include illustrations of their secondary structures, Gibbs free energies, simulation results, molecular dynamics simulation videos and more. The database is freely accessible at: www.sbi.uni-rostock.de/triplexrna.

The following section is to motivate and discuss the approaches that we integrated and the tools and algorithms that we used in our study. Thereafter, our case study and the results obtained are described in detail. Finally, the possible impact of our findings is set into in a wider context and possible avenues for further investigation are discussed.

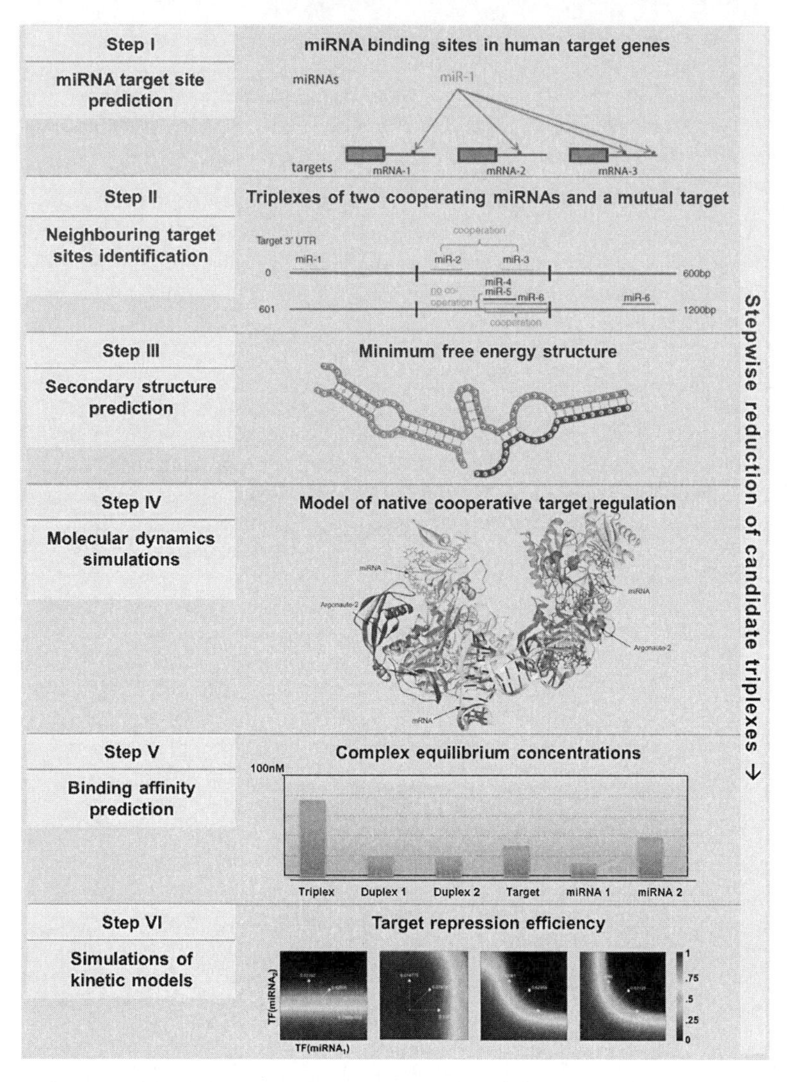

Figure 4.2 Schematic representation of the PreCoM workflow.
The workflow is comprised of six steps (left) that integrate methods from different disciplines for an agglomeration of *in silico* evidence. Next to each step the expected result is illustrated (right). The set of candidate triplexes shrinks as the workflow advances because in each step putatively non-functional RNA triplexes are discarded.

4.2 Research strategy

4.2.1 Predicted and validated miRNA target interactions

The first step of the PreCoM workflow, i.e. the identification of miRNA target sites, lays the foundation for all the subsequent predictive and analytical steps. A number of targets of miRNA regulation have already been experimentally identified by RNA-based, immunoprecipitation-based, or protein-based approaches (Martinez-Sanchez and Murphy, 2013; Bhattacharya and Kunz, 2013). However, in many cases neither the direct interaction of miRNAs with their target mRNA was proven, nor the exact binding site determined. A few web-accessible databases exist, e.g. TarBase (Papadopoulos *et al.*, 2009), miRTarBase (Hsu *et al.*, 2011) and miRecords (Xiao *et al.*, 2009b), that accumulate data on miRNA-target interactions with experimental support from reports in the scientific literature. We have reviewed these and other miRNA-based web resources in (Schmitz and Wolkenhauer, 2013). In most studies, miRNA induced target repression is confirmed by qRT-PCR, microarray, western blot, pSILAC, immunoprecipitation or luciferase assays, where the last method provides the strongest evidence for a functional miRNA-target interaction. Nevertheless, only in rare cases evidence is provided for the specific binding site (e.g. through site-directed mutagenesis). A recently developed experimental technique for the identification of exact miRNA binding sites, CLASH (crosslinking, ligation, and sequencing of hybrids; Kudla *et al.*, 2011), may change this situation. CLASH has already substantiated a large number of predicted binding sites in a high-throughput experiment in human embryonic kidney (HEK) cells conducted by Helwak and colleagues (Helwak *et al.*, 2013). However, due to the lack of evidence for the specific binding sites in most validated miRNA-target interactions, we suggest the use of prediction algorithms for an *in silico* identification of exact binding sites.

A plethora of miRNA-target prediction algorithms have been designed in the last decade, some of which are better in terms of sensitivity (true positive rate) and others in terms of specificity (true negative rate). The first group tends to have a higher fall out (false positive rate), while the latter achieves specificity at the cost of sensitivity, i.e. overall less valid miRNA-target interactions are predicted. See Chapter 2 for a discussion of different approaches used in miRNA-target prediction algorithms. A good compromise between sensitivity and specificity is given in the case of the miRanda algorithm (Betel *et al.*, 2010).

miRanda was one of the first developed miRNA-target prediction algorithms (Enright *et al.*, 2003). It uses a modified Smith-Waterman algorithm (a position weighted local alignment algorithm) in that complementarity towards the 5′ end of the miRNA is rewarded (higher scoring factor for the first 11 positions) and position specific empirically defined rules are applied. Additionally, miRanda identifies a target site as being conserved if a miRNA independently matches an orthologous site in the same position in a related species. Furthermore, structure prediction and thermal stability calculation are used for scoring putative miRNA-target pairs. Recently, a new scoring system for predicting the likelihood of target repression by a miRNA was proposed (Betel *et al.*, 2010). The mirSVR score, which is applied in miRanda predictions, is based on a support vector regression classifier that was trained with data from miRNA transfection experiments. Sequence, contextual and structural features of predicted miRNA target sites taken from down-regulated target mRNAs were incorporated into this classification system where scores are correlating with the extent of target down-regulation.

miRanda predictions largely overlap with those from all other algorithms (highest relative overlap) (Enright *et al.* 2003; Ritchie *et al.* 2009). Therefore, and because we wanted a good coverage (sensitivity) but not too many false positive predictions, we chose the miRanda algorithm for our study.

Seed site distance

In line with the theory proposed by Saetrom and colleagues (2007), stating that miRNA pairs whose binding sites reside in close proximity (13-35nt seed site distance) in a mutual target gene achieve an enhanced repression efficiency, we examined all miRanda predicted binding sites in mRNA 3′ UTR sequences for such instances using a customized Python script.

3′ UTR sequences of many genes exhibit several, sometimes dozens of miRNA target sites. Therefore, many instances of neighbouring binding sites can be found that fulfil the before mentioned prerequisite for cooperative miRNA regulation. The pressing question is, whether all the associated miRNA pairs are able to cooperate and if not, how to identify and eliminate false candidates. To this end, we suggest a more in depth analysis of RNA triplexes in order to minimize false positive predictions. The coming steps in the PreCoM workflow can be used on the one hand to increase confidence in selected candidates and on the other hand to discard candidates that are likely to be non-functional. Ultimately, the proposed analyses reduce the risk of failure in subsequent validation experiments.

4.2.2 Secondary structure and minimum free energy prediction

A widely established method in miRNA-target prediction algorithms is the prediction of the minimum free energy structure of a putative RNA hybrid formed by the miRNA and its target (Rehmsmeier *et al.* 2004; Rajewsky 2006). The free energy is an accepted measure for the thermodynamic stability of a RNA structure composed of one or more sequences. Therefore, we suggest local secondary structure prediction of candidate triplexes as the third step in the PreCoM workflow (Figure 4.3). One hypothesis is that in the case of a stable RNA triplex, more effective target repression can be achieved. Furthermore, we were wondering if we can find structural features correlating with the target repression efficiency.

One factor that is crucial, at least for functional miRNA-target interactions, is the seed binding (Wang and El Naqa 2008; Gan and Gunsalus 2013). The miRanda algorithm itself considers perfect seed binding as a minimum requirement in predicted miRNA-target pairs. We therefore also suggest looking for preserved seed binding of both miRNAs in the predicted secondary structure of RNA triplexes.

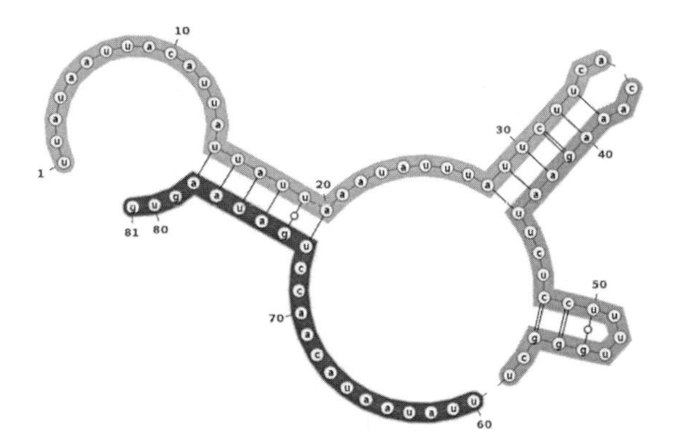

Figure 4.3 Local secondary structure of a RNA triplex.
This illustration shows is the minimum free energy structure of the RNA triplex composed of the miRNAs miR-186 and miR-374a, as well as a subsequence of their mutual target mRNA (*ZCCHC4*) that contains both miRNA binding sites and the connecting nucleotides. Of note, in this example seed binding is preserved for only one of the two miRNAs. Therefore, in case of a stringent selection approach, this triplex would have to be discarded from the candidate set. For visualization we used the RNA structure drawing tool VARNA (version 3.9; Darty *et al.*, 2009).

For the prediction of the triplex secondary structure we retrieved gene 3′ UTR sequences from the RefSeq gene track of the UCSC table browser (GRCh37/hg19; Meyer *et al.* 2012) and mature miRNA sequences from the miRBase database (release 20; Griffiths-Jones *et al.*, 2006). For structure prediction we used the *mfe* tool of the NUPACK software package (Zadeh *et al.*, 2011). Each triplex structure includes the mature sequences of both putatively cooperating miRNAs and the mRNA subsequence that includes both predicted miRNA binding sites. *mfe* computes the full partition function (except for pseudo knot structures) of RNA complexes in dilute solution and determines the minimum free energy structure, which is returned in dot-bracket notation together with an estimate of the triplex free energy (TFE; a.k.a. Gibbs energy change or ΔG).

4.2.3 Tertiary structure modelling and molecular dynamics simulations

In secondary structure prediction, the goal is to identify the minimum free energy structure of a nucleic acid sequence or complex, i.e. based on the assumption that stacking base pairs and loop entropies contribute additively to the energy of a nucleic acid secondary structure (Mathews *et al.*, 1999a; Mathews, 2004), one or more structures with minimum free energies can be computed. However, a more realistic representation is achieved by a tertiary (3D) structure model. The thermodynamic stability is *in silico* best determined through MDS (McDowell *et al.*, 2007). Therefore, we proposed as the fourth step in the PreCoM workflow the prediction of the 3D structure of RNA triplexes and the derivation of a thermodynamic profile by performing MDS. The thermodynamic profile is one pre-requisite for the prediction of target repression efficiencies and the synergistic effect achieved by cooperatively acting miRNAs.

Three-dimensional model construction

Template-based and template-free methods for RNA 3D structure modelling already exist (Rother *et al.*, 2011). However, there is no algorithm available for the prediction of RNA complex 3D structures. We used RNAComposer (Popenda *et al.*, 2012), a fully automated tool for constructing large RNA 3D structures from user provided secondary structure information, to predict tertiary structures of RNA triplexes as well as of their inherent duplexes (mRNA+miR_1 and mRNA+miR_2) based on the secondary structures computed in the previous step. RNAcomposer translates RNA secondary structures by using a library (RNA FRABASE database; Popenda *et al.*, 2010) that links RNA secondary structure and tertiary structure elements based on data from the Protein Data Bank (PDB; Berman, 2000). RNAComposer is designed to predict 3D structures for large single-stranded RNA sequences; therefore we had to concatenate the RNA sequences involved in the RNA triplex, an approach commonly used in RNA complex structure prediction methods (Paciello *et al.*, 2011; Gan and Gunsalus, 2013; Figure 4.4A). After 3D structure prediction we had to delete the phosphodiester bonds that connect the mRNA 3′ terminal base with the $miRNA_1$ 5′ terminal base, as well as the $miRNA_1$ 3′ terminal base with the $miRNA_2$ 5′ terminal base. Disjoining of the RNA termini was performed using the Build and Edit Nucleic Acid tool in Accelrys® Discovery Studio 3.5. To avoid re-joining, we changed the phosphate group at the 5′ end of each RNA species to a hydroxyl group using the CapNucleotide function.

We subsequently optimized the geometry of the 3D model using an energy minimization protocol in Accelrys® Discovery Studio 3.5 which has been successfully applied before (Hart *et al.*, 2005). For details see Appendix D.1.

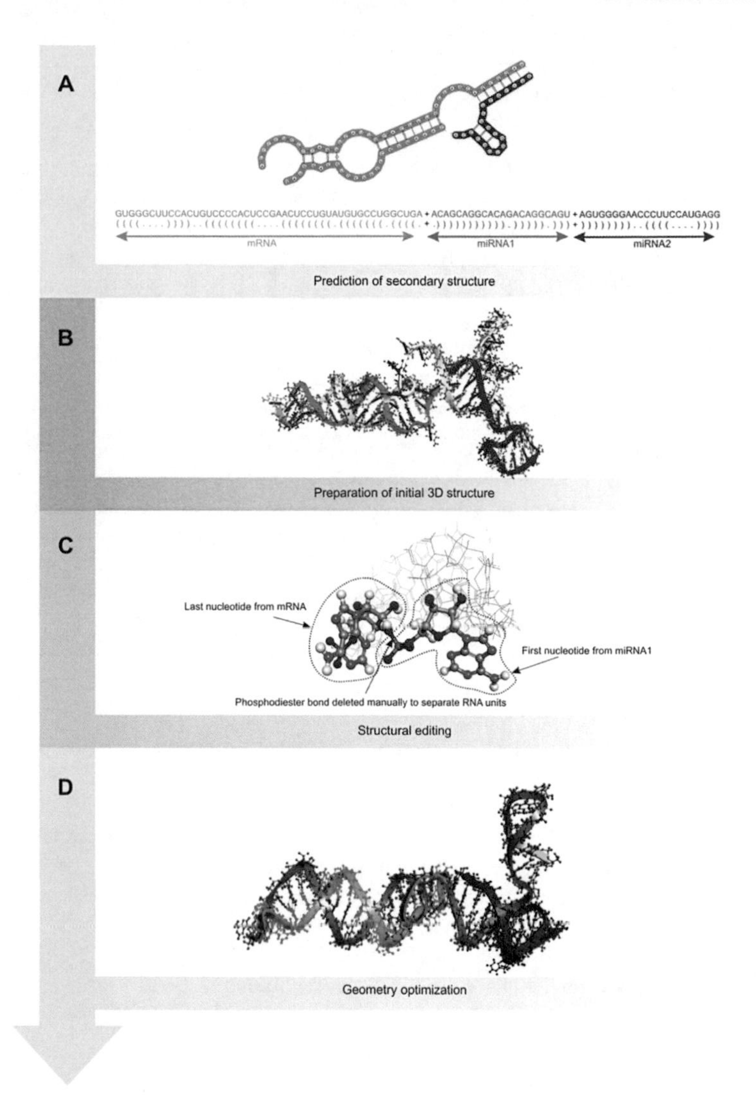

Figure 4.4 Workflow for the construction of tertiary RNA triplex structures.
(**A**) Prediction of the RNA triplex secondary structure, illustrated here as 2D structure and in dot-bracket notation. The molecules involved are color-coded: target mRNA (grey), $miRNA_1$ (green) and $miRNA_2$ (red). (**B**) Prediction of the 3D structure of the RNA triplex. In the depicted initial 3D structure all RNA strands are concatenated due to the fact that no algorithm for direct 3D structure prediction of RNA complexes exists. All RNA strands were concatenated to mimic a long single RNA strand. (**C**) Structure editing: (i) Deletion of phophodiester bonds that connect the mRNA 3′ terminal base with the $miRNA_1$ 5′ terminal base, as well as the $miRNA_1$ 3′ terminal base with the $miRNA_2$ 5′ terminal base. (ii) Capping of the first nucleotide (5′ terminal) in all RNA species. (**D**) Final 3D model of the RNA triplex after geometry optimization.

Molecular dynamics simulations

The optimized structure of the complexes was gradually heated from 50 K to 300 K by scaling the velocity of each atom in a total of 10,000 steps with 1 fs/step. After this heating phase, the system was stabilized around the target temperature of 300 K by performing equilibration. RNA complexes were verified for their stability after the equilibration phase. For stable complexes, MD production simulations were performed, initially 100 ps duration. The production phase for stable complexes after 100 ps of initial run was extended for another 400 ps to check the stability of the RNA complexes. For more details on the molecular dynamics simulations see Appendix D.1.

4.2.4 Prediction of complex equilibrium concentrations

In the fifth step of the PreCoM workflow, we suggest a partition function algorithm to compute the equilibrium probability distribution of complexes (duplexes, triplexes etc.) and monomers that can arise from an RNA triple including two miRNAs and their mutual target mRNA (Dirks *et al.*, 2007). Such a partition function algorithm is provided as part of the NUPACK package, a software collection for the analysis and design of nucleic acid systems (Zadeh *et al.*, 2011). Based on given initial concentrations, using the tools *complexes* and *concentrations* from the NUPACK package, we can derive equilibrium concentrations for each complex species in a fixed volume of dilute solution. We propose to assign initial concentrations of 100 nM (nanomolar) to all RNA species which reflects typical concentrations used in RNA transfection experiments. However, the retrieved equilibrium concentrations, which are representative for RNA binding affinities, provide an estimate for the ratios of concentrations among the complex species and thus reveal propensities toward triplex formation. Predominant formation of RNA triplexes may be the basis for cooperative miRNA target regulation. Furthermore, we propose to derive putative RNA association rates from the complex equilibrium concentrations, which are used in the next step of the PreCoM workflow as parameter in the model of synergistic target regulation.

4.2.5 A mechanistic model of synergistic target regulation

In the previous steps we suggested the sequence-based prediction of putatively cooperating miRNAs and their mutual target, building 2D and 3D models of RNA triplexes and determining their thermodynamic stability, and the prediction of equilibrium concentrations of the RNA complex species. In the last step of the PreCoM workflow we suggest to assess the target repression efficiency and the synergistic effect attained by cooperating miRNAs. The most appropriate tool to achieve this goal is a mechanistic model composed of ODEs that can be used to simulate target steady states or in general the temporal evolution of the molecular entities in a biological system (Lai *et al.*, 2013). As already discussed in Chapter 3, such a model describes possible reactions concerning the involved molecules, i.e. synthesis, complex formation and dissociations as well as degradation. See Figure 4.5 for an illustration of the model structure in SBGN format.

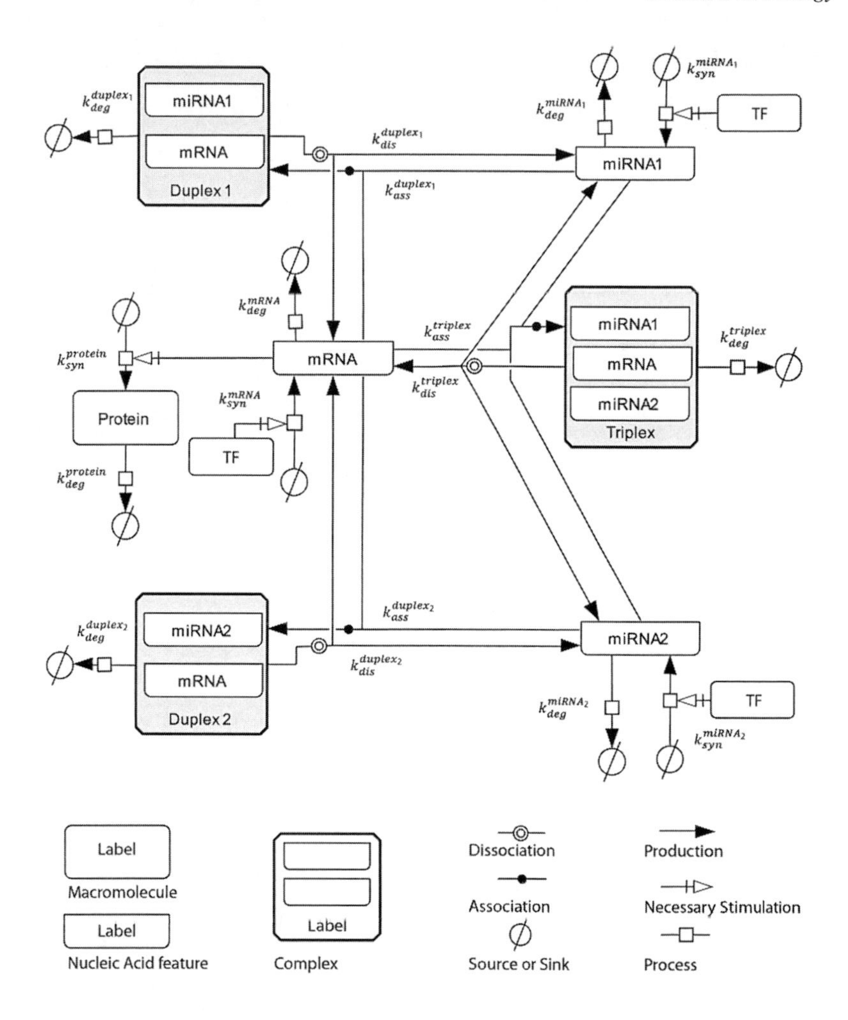

Figure 4.5 Process diagram of the mechanistic model of cooperative target regulation.
The biochemical reaction network underlying the phenomenon of cooperative target regulation is composed of two miRNA species and the target mRNA and protein. For all molecules we consider synthesis and degradation processes. Furthermore, we consider the formation of miRNA-mRNA duplexes and a triplexes composed of two miRNAs and the target mRNA. All complexes depend on association/dissociation and degradation processes. The process diagram is depicted in a standaradized way (according to the SBGN). Labels of the connecting arrows correspond to the names of the rate constants in the ODE system.

The proposed model for synergistic target regulation consists of the following equations:

$$\frac{dprotein}{dt} = k_{syn}^{protein} \cdot mRNA - k_{deg}^{protein} \cdot protein$$

$$\frac{dmRNA}{dt} = k_{syn}^{mRNA} \cdot TF_{mRNA} - k_{deg}^{mRNA} . mRNA - k_{ass}^{duplex_1} \cdot mRNA \cdot miRNA_1 - k_{ass}^{duplex_2} \cdot mRNA \cdot miRNA_2 - k_{ass}^{triplex} \cdot mRNA \cdot miRNA_1 \cdot miRNA_2 + k_{dis}^{duplex_1} \cdot duplex_1 + k_{dis}^{duplex_2} \cdot duplex_2 + k_{dis}^{triplex} \cdot triplex$$

$$\frac{dmiRNA_1}{dt} = k_{syn}^{miRNA_1} \cdot TF_{miRNA_1} - k_{deg}^{miRNA_1} \cdot miRNA_1 - k_{ass}^{duplex_1} \cdot mRNA \cdot miRNA_1 - k_{ass}^{triplex} \cdot mRNA \cdot miRNA_1 \cdot miRNA_2 + k_{dis}^{duplex_1} \cdot duplex_1 + k_{dis}^{triplex} \cdot triplex$$

$$\frac{dmiRNA_2}{dt} = k_{syn}^{miRNA_2} \cdot TF_{miRNA_2} - k_{deg}^{miRNA_2} \cdot miRNA_2 - k_{ass}^{duplex_2} \cdot mRNA \cdot miRNA_2 - k_{ass}^{triplex} \cdot mRNA \cdot miRNA_1 \cdot miRNA_2 + k_{dis}^{duplex_2} \cdot duplex_2 + k_{dis}^{triplex} \cdot triplex$$

$$\frac{dduplex_1}{dt} = k_{ass}^{complex_1} \cdot mRNA \cdot miRNA_1 - k_{deg}^{duplex_1} \cdot duplex_1 - k_{dis}^{duplex_1} \cdot duplex_1$$

$$\frac{dduplex_2}{dt} = k_{ass}^{duplex_2} \cdot mRNA \cdot miRNA_2 - k_{deg}^{duplex_2} \cdot duplex_2 - k_{dis}^{duplex_2} \cdot duplex_2$$

$$\frac{dtriplex}{dt} = k_{ass}^{triplex} \cdot mRNA \cdot miRNA_1 \cdot miRNA_2 - k_{deg}^{triplex} \cdot triplex - k_{dis}^{triplex} \cdot triplex$$

The model contains ODEs describing the temporal evolution of monomers and complex species involved in cooperative target regulation, i.e. two miRNAs, the target mRNA and protein, possible miRNA-mRNA duplexes, and the triplex (two miRNAs + one mRNA). The biosynthesis of all polynucleotides ($mRNA$ and $miRNA_i$) depends on the presence of their corresponding TF (respectively, TF_{mRNA} and TF_{miRNAi}) and their individual synthesis rate ($k_{syn}^{<mRNA,miRNA_i>}$). The biosynthesis of the target protein depends on its basal synthesis rate ($k_{syn}^{protein}$) and the presence of free target mRNA. The RNA molecules, the complexes and the target protein are also subject to degradation processes reflected by the degradation rate constants ($k_{deg}^{<mRNA,miRNA,duplex,triplex,protein>}$). The concentrations of free miRNAs and mRNA molecules decrease upon duplex and triplex formation (k_{ass}^{duplex}; $k_{ass}^{triplex}$) and increase when complexes are dissociated into their constituent parts ($k_{dis}^{<duplex,triplex>}$).

For the parameterization of the model we propose the following strategy: For the association and dissociation rate constants we utilize the equilibrium probability distribution derived from step five in the PreCoM workflow and the potential energy (PE) values derived from step four. There is no experimental data on RNA complex' association and dissociation velocities available. Therefore, we consider the predicted equilibrium concentrations, reflecting probabilities for complex formation, as suitable for characterizing the association rate constants ($k_{ass}^{<duplex_i,triplex>}$). Furthermore, we consider PE values, which reflect the thermodynamic stability of a complex, as suitable for parameterizing the dissociation rate constants ($k_{dis}^{<duplex_i,triplex>}$). We make these parameters comparable by normalizing them to values in the range between 0 and 1 (Figure 4.6).

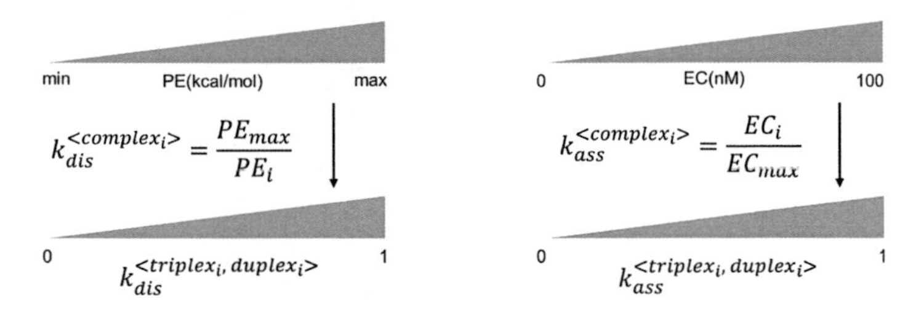

Figure 4.6 Parameter characterization and normalization.
We characterized the association and dissociation rate constants using the potential energies (PE) of the complexes and their equilibrium concentrations (EC). The characterized parameter values were normalized using the depicted equations (middle), which ensures that the obtained parameter values for the association and disassociation rate constants of the complexes are constrained in the range [0 1].

For modelling the target repression efficiency we suggest to fix the synthesis and degradation rate constants of the molecular species involved to the value 1. Thereby, the nominal steady state of all involved RNA species equals 1. Now, the protein expression level clearly indicates if translation repression occurs, i.e. in case the steady state value is < 1. In case the protein steady state value is ≥ 1, miRNAs do not repress translation.

Using the model, we suggest simulating target protein steady states for varying concentrations of the cooperating miRNAs. More specifically, we propose a range $[10^{-1}\ 10^{2}]$ for the TF-dependent miRNA expression (TF_{miRNAi}), representing down- and up-regulation of the miRNAs, in which, for pre-defined intervals, different combinations of miRNA expression are applied and the target protein steady state is computed. The hypothetical synergy of two putatively cooperating miRNAs can be made visible with the help of a contour plot (Figure 4.7). The possible evidence for synergistic target regulation can be substantiated by introducing reference points representing characteristic miRNA expression values, e.g. (a) normal expression (1,1); (b)+(c) overexpression of $miRNA_1$ (10,1) or $miRNA_2$ (1,10); and (d) moderate up-regulation of both miRNAs (5,5). If the target repression in reference point (d) is beyond the expected additive effect achieved by combinatorial target regulation (compared to reference points (b) and (c)), a case of synergistic target regulation can be assumed. Here we introduce a measure for the repression gain achieved by single ($RG_{<1,2>}$) or combinatorial (RG_3) miRNA up-regulation. The repression gain is calculated by subtracting the target steady state under normal conditions from that of miRNA up-regulation ($RG_i = SS_{up} - SS_{normal}$). The higher the RG the stronger the repression effect on the target gene. Synergistic target regulation is realized when $RG_3 > RG_1 + RG_2$.

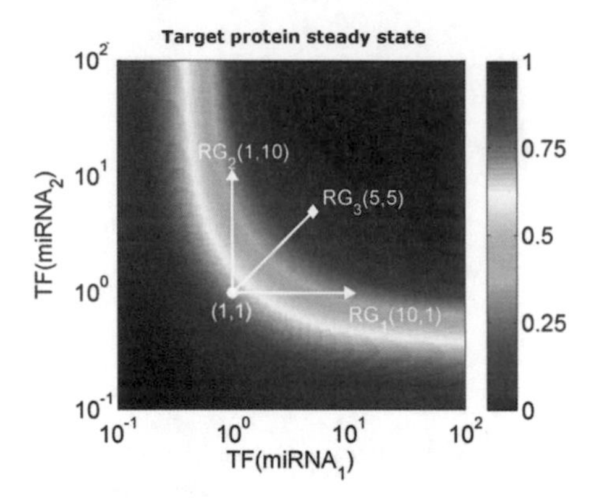

Figure 4.7 Target steady states for varying expression rates of cooperating miRNAs. Transcription factor-dependent miRNA expression values ($TF(miRNA_i)$) are applied in the range $[10^{-1}\ 10^2]$. The simulated target protein steady states are visualized using a rainbow colour scale (see legend) ranging from red (high concentration) to blue (low concentration). Due to the simplification mentioned above, the protein expression levels computed are constrained between 0 (target is silenced) and 1 (no repression). The plot includes four reference points representing characteristic miRNA expression values: (a) normal expression $(1,1)$; (b)+(c) overexpression of miRNA$_1$ $(10,1)$ or miRNA$_2$ $(1,10)$; and (d) moderate up-regulation of both miRNAs $(5,5)$. The repression gain (RG) is determined for the scenarios of single miRNA overexpression (RG_1 and RG_2) or combined up-regulation of both miRNAs (RG_3). Synergistic target regulation is realized when $RG_3 > RG_1 + RG_2$.

4.2.6 Data sharing

By implementing and conducting our proposed workflow for the identification and analysis of pairs of cooperating miRNAs and their mutual targets, a set of heterogeneous data is generated. The data collected and produced consists of: (1) miRNA and target sequences (including miRNA target site coordinates); (2) secondary structures (as image and in dot-bracket notation; plus values for the Gibbs energy change (TFE); (3) 3D triplex structures (as image and as Protein Data Bank (PDB) file including the atomic coordinates); (4) molecular dynamics simulation movies (including energy profile and stability time); and (5) kinetic models and simulation experiments (including parameter values and contour plots). Analytical methods such as those proposed in the following paragraphs will generate more metadata and annotations for the primary data summarized above. No standardized format exists in which these data as a whole can be archived and shared with the community. Therefore, and due to the fact that there is no database which integrates the different data types generated in the PreCoM workflow, we decided to implement our own dedicated database. We established a relational database named TriplexRNA using MySQL server (version 5.5.22), which is equipped with a web interface realized through CGI scripts (written in Python v2.7.2). This facilitates easy access for the users. Furthermore, we provide a programmable interface to the database using the representational state transfer (REST) architecture. This was realized using the Bottle web framework for Python (http://bottlepy.org).

4.3 Case study

We implemented the PreCoM workflow in a case study in order to predict and analyse putative human target genes of cooperative miRNA regulation. It should be noted that the workflow can be applied in any other species in the animal kingdom because miRNA genes as well as their functional mechanisms are widely conserved among animals (Bartel, 2004). Details about the analysis of human RNA triplexes are described in detail in the following paragraphs.

4.3.1 Identification of miRNA target sites

The human genome contains an estimated 20,389 protein-coding genes (GRCh38 - The Genome Reference Consortium). Many of these genes are post-transcriptionally regulated by miRNAs (Chang and Mendell, 2007). We based our analysis of cooperating miRNAs on predicted miRNA binding sites in human protein-coding genes. More specifically, we considered pairs of miRNAs as putative cooperation partners when their binding sites reside in close proximity (13-35nt seed site distance) in the 3′ UTR of a mutual target mRNA. In order to identify such instances, we retrieved data on predicted miRNA-target interactions from the microrna.org web repository, which provides results based on the miRanda target prediction algorithm (Betel *et al.* 2010; John *et al.* 2004). Only those target sites of conserved miRNAs and with good prediction scores ($S_{miRSVR} \leq -0.1$) were considered.

Information on experimentally supported miRNA target interactions was derived from the miRTarBase (release 4.5; Hsu *et al.*, 2011). miRTarBase contains data that were manually extracted from the literature. It is currently the largest resource of its kind. Target sites for validated miRNA-target interactions (extracted from the afore mentioned resources) were also derived from miRanda predictions.

In total we identified 1,097,065 putative miRNA binding sites in the 3′ UTR sequences of human protein coding genes.

4.3.2 Cooperative target regulation may effect most human protein coding genes

Based on the seed site distance constraint (13-35 nt), we identified 17,259 human genes as putative targets of cooperative miRNA regulation. This accounts for more than 4/5th of all protein coding genes (~85%). According to these results, most of the target genes are post-transcriptionally regulated by a small number of miRNA pairs, but there are also many cases of highly regulated target hub genes that may be cooperatively regulated by dozens or even hundreds of miRNA couples. Of note, we did not observe any preferential seed site distance among the identified target site pairs, except for a frequency decline towards larger distances. MiRNA target hub regulation was discussed and analysed in detail in the previous chapter and in Lai *et al.* (2012). The histogram depicted in Figure 4.8 illustrates the number of genes predicted to be targeted by an ascending number of cooperating miRNA pairs. However, frequencies for large numbers of miRNA pairs per gene drop when constraints beyond the seed site vicinity are applied (e.g. for the free energy of the triplex or the triplex' equilibrium concentration). These constraints are discussed in detail in the following paragraphs.

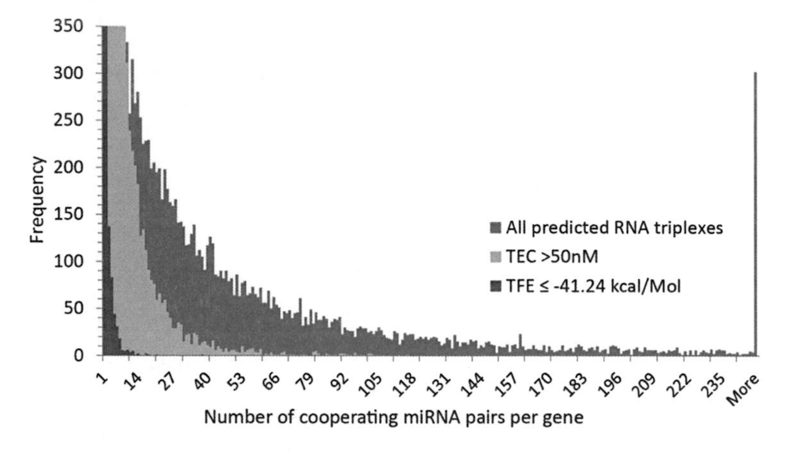

Figure 4.8 Number of genes targeted by an ascending number of cooperating miRNA pairs. Most human genes are targeted by less than a dozen cooperating miRNA pairs. However, there seems to be a number of highly regulated target hub genes with several dozen or even hundreds of binding sites for cooperating miRNAs (blue bars; y-axis has been truncated; frequencies for values > 250 have been added up at the right-most tick mark on the x-axis). These results are based on miRanda predicted miRNA target sites in mRNA 3′ UTRs of human protein coding genes and the necessary seed site distance of 13-53nt between two neighbouring binding sites. Frequencies drop when constraints concerning the binding affinity of RNA species (e.g., predicted $TEC > 50\,nM$, green bars) and the stability of RNA triplexes (e.g., predicted $TFE \leq -41.24\,kcal/mol$, red bars) are applied.

4.3.3 Triplexes with conserved target sites have lower binding affinities

Nucleic acid sequences that are conserved across species are thought to have functional value. In the case of miRNA regulation, there is evidence that targets with conserved seed sites are more strongly repressed by miRNAs compared to targets with non- or weakly conserved seed sites (Selbach *et al.*, 2008). In line with this, Friedman and colleagues found in miRNA transfection experiments that down-regulated targets with conserved miRNA binding sites are in the majority (Friedman *et al.*, 2009). Based on these studies, we aimed to determine the importance of conservation in cooperative target regulation. However, as we lack data from large scale experiments determining the impact of cooperating miRNAs on target repression, we instead inspected the predicted TECs in order to derive possible conservation-based dependencies for efficient triplex formation (Figure 4.9).

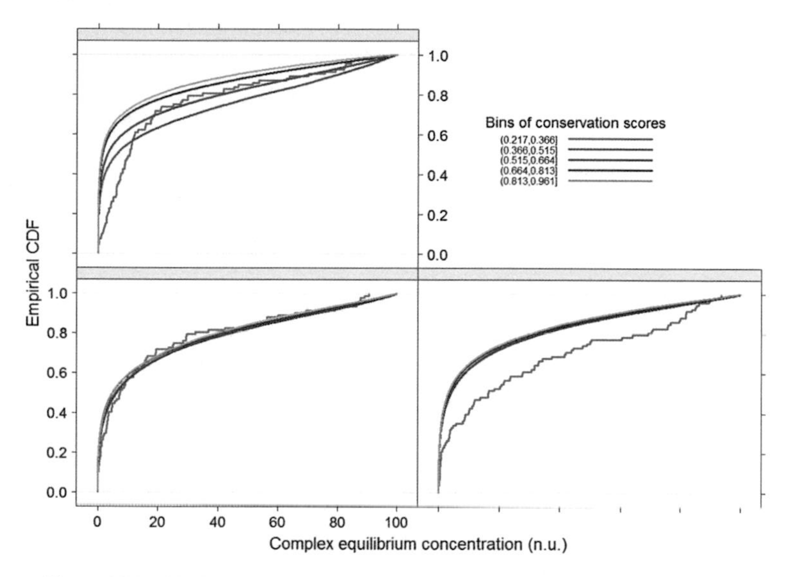

Figure 4.9 Equilibrium concentrations of RNA triplexes and inherent duplexes.
The three plots illustrate the empirical cumulative distribution functions (ECDF) of predicted triplex and duplex equilibrium concentrations (binding affinities) for different degrees of target site conservation (PhastCons score; Siepel *et al.*, 2005) distributed over five bins as defined in the figure legend. Conservation scores range between 0 (no conservation) and 1 (perfect conservation). The number of triplexes and duplexes with the least conserved target sites (blue curves) is too small and was therefore not included in the interpretation of the results. **Top:** ECDF of triplex equilibrium concentrations. Triplexes with strongly conserved target sites tend towards lower equilibrium concentrations (yellow and red curves), while triplexes with poorly conserved target sites have higher concentrations by trend (green and pink curves). **Bottom:** ECDF of miRNA-mRNA duplex equilibrium concentrations (left and right). In contrast to the triplexes, duplex equilibrium concentrations are equally distributed independent of the target site conservation.

Unexpectedly, we found equilibrium concentrations of triplexes with strongly conserved target sites to be generally lower than those of triplexes with weakly conserved target sites (high TECs indicate a preference towards triplex formation). Interestingly, this observation cannot be made for the inherent miRNA-mRNA duplexes. These results can be interpreted as follows: Based on the studies conducted by Selbach *et al.* (2008) and Friedman *et al.* (2009) we can assume, on the one hand, that miRNAs are effective regulators of target genes in which they can bind to well conserved target sites. Poorly conserved target sites, on the other hand, are often not functional and may depend on the support of a second proximate miRNA target site. In this case target repression may be enabled or enhanced by stabilization through RNA triplex formation. A detailed analysis of this hypothesis is described in the following paragraphs.

For now, we can conclude that by considering the seed site distance of predicted miRNA target site pairs and their conservation in other species alone, one cannot differentiate between functional and non-functional RNA triplexes. In case of the seed site distance constraint, miRNAs are predicted to cooperate with almost all (sometimes all) other conserved miRNAs in the regulation of one or the other target mRNA. Therefore, and in order to narrow down predictions to a set of high-confidence candidates of functional RNA triplexes, we investigated the structural conformations and thermodynamic stabilities as well as RNA complex binding affinities in the course of our case study. These and other analyses are described below.

4.3.4 Canonical RNA triplexes form the thermodynamically most stable structures

Secondary structures of functional RNA molecules are more stable than random RNA structures (Rivas and Eddy, 2000). Consequently, we determined the local secondary structure and the thermodynamic stability of the predicted RNA triplexes. Furthermore, we analysed the results for favoured and beneficial conformational and architectural patterns. We then used this information to segregate non-functional from functional RNA triplexes.

Secondary structure prediction of RNA triplexes revealed four different classes of local structural conformations: (i) canonical triplex (CT); (ii) triplex with target self-complementarity (TSC), i.e. intra-molecular base pairing in the target sequence; (iii) triplex with miRNA self-complementarity (MSC), i.e. intra-molecular base pairing in at least one miRNA sequence; and (iv) triplex involving miRNA-miRNA hybridization (MD), i.e. inter-molecular base paring between both involved miRNAs. The four structural conformations are depicted in Figure 4.10.

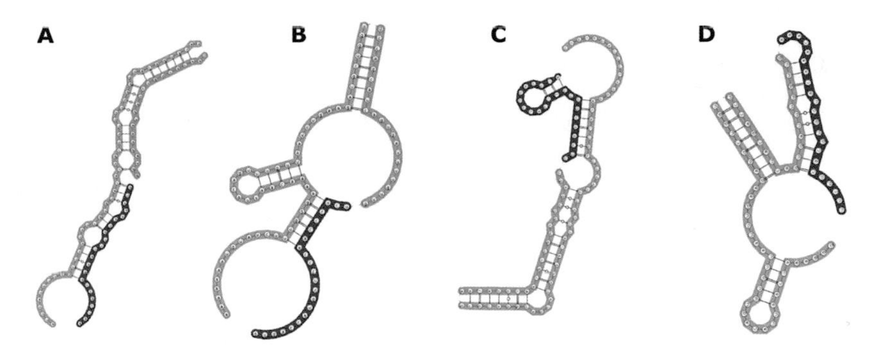

Figure 4.10 Four structural conformations of 2D RNA triplex structures.
In the four sketches, the target mRNA is coloured in grey while the two cooperating miRNAs are coloured in red and green respectively. (**A**) canonical triplex (no intra- nor inter-molecular base pairing except for the expected hybridization between the miRNAs and their designated binding sites); (**B**) triplex with target self-complementarity (intra-molecular base pairing in the target sequence); (**C**) triplex with miRNA self-complementarity (intra-molecular base pairing in at least one miRNA sequence); and (**D**) triplex involving miRNA-miRNA hybridization (inter-molecular base paring between both involved miRNAs).

Sometimes minimum free energy structure prediction of a RNA triple suggests a RNA duplex and one isolated RNA molecule as optimal structural arrangement. Cooperative miRNA regulation cannot take place in these cases. We therefore discarded these RNA triples form further analysis. The other conformations mentioned and shown in Figure 4.10 were further analysed as described below.

Muckstein *et al.* (2006) showed that the efficiency of RNA interference correlates with the binding energy between siRNA (or miRNA) and target mRNA. This suggests that in the case of RNA triplexes, the binding energy may also be crucial for efficient target repression and the synergistic effect achieved through cooperative miRNA regulation. We therefore computed the triplex free energies (TFEs), in other words Gibbs free energies ($\Delta G_{triplex}$), of the predicted RNA triplexes. Compared to the Gibbs free energies of the intrinsic miRNA-mRNA duplexes the TFE value is typically much lower. We found only one case (a canonical triplex composed of miR-137, miR-374a and *MYB*) out of 674,310 predicted RNA triplexes in which the Gibbs free energy ($\Delta G_{duplex} = -9.76$) of one of the intrinsic duplexes was slightly lower than the TFE ($\Delta G_{triplex} = -9.56$). In all other cases triplex formation was beneficial in terms of free energy. One can utilize this measure to short-list possible RNA triplexes, considering only those as high-confidence candidates which stay below a certain cut-off value. For example, when we apply a cut-off at $\Delta G_{triplex} = -41.24\ kcal/mol$, the number of target genes of cooperative miRNA regulation drops from an initial 17,259 to 1,779. See also Figure 4.8 for how the number of cooperating miRNA pairs per gene changes when this cut-off is applied. Likewise, the number of cooperation partners per miRNA drops from an initial 242.29 to 23.0. The proposed cut-off value is three standard deviations below the mean TFE value of all predicted RNA triplexes ($Z = 3$). This Z-value was previously proposed by Stark *et al.* (2003) for the identification of high-confidence miRNA-target duplexes.

Next, we wanted to determine the thermodynamic benefit of forming a triplex over a duplex structure. Therefore, we introduced a new measure to express this benefit which we refer to as energy gain ($\Delta\Delta G$). The energy gain is determined by subtracting the lower free energy value of the two possible duplexes from the TFE value ($\Delta\Delta G = \Delta G_{triplex} - \Delta G_{duplex\_min}$). In our case study of human RNA triplexes, energy gain values ranged between $0.22\ kcal/mol$ and $-24.08\ kcal/mol$. We observed that the formation of canonical triplexes typically releases more energy in comparison to other structural conformations, i.e. more energy is gained in this case (Figure 4.11, right). Their TFE values and those of TSCs are in general lower compared to MSC and MD structures (Figure 4.11, left). Based on these results, we propose the energy gain value ($\Delta\Delta G$) as another mean for the identification of effective and efficient RNA triplexes.

Until this point we did not look at the seed binding of the miRNAs to the target mRNA. It was shown by Wang and El Naqa (2007) and Gan and Gunsalus (2013) that seed binding is crucial for efficient target repression. In the class of RNA triplex structures involving miRNA-miRNA hybridization we observed that the seed binding for at least one of the two miRNAs is lost, i.e. the 6-8 nt region at the 5′ end of one or both miRNAs does not hybridize with the target. One such example is illustrated in Figure 4.10D. In the other structural triplex conformations one can also observe cases were miRNA seed bindings are not preserved (precisely in 38.38% of all cases). Of note, in the case of the above mentioned RNA triple for which duplex formation is thermodynamically favourable compared to triplex formation, seed binding for both miRNAs is not preserved in the secondary structure. Based on this, and the observations made by Wang and El Naqa (2007) as well as Gan and Gunsalus (2013), we can assume that such triplexes are non-efficient. For this reason we discarded them from further analyses. In conclusion we can assume that the structural conformation of RNA triplexes is linked to their thermodynamic stability and therefore also to the target repression efficiency.

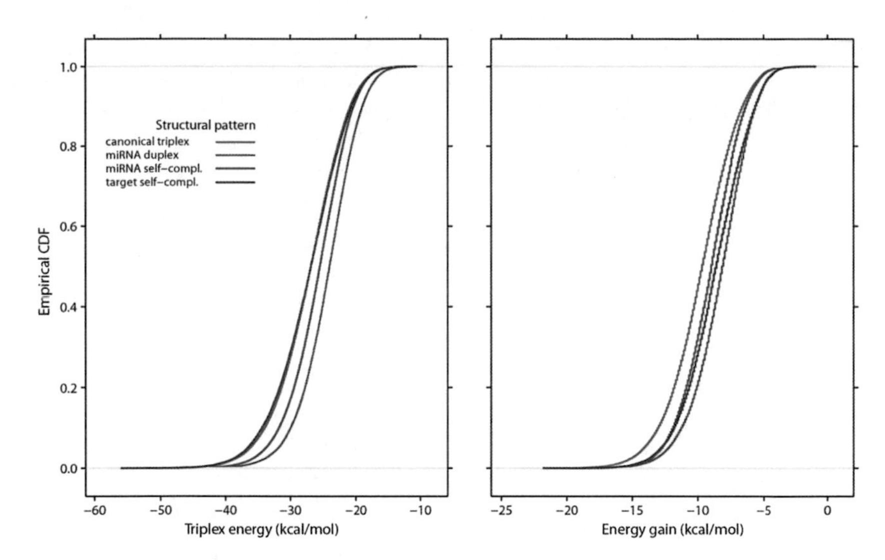

Figure 4.11 Cumulative distributions of triplex free energies and free energy gains.
Left: Lower TFE values can be observed for canonical triplexes (blue line) and triplexes with target self-complementarity (red line). This suggests stronger stability in these structural conformations. **Right:** When we compare the empirical cumulative distribution functions of all structural conformations in terms of energy gain, canonical triplex formation turns out to be most beneficial.

4.3.5 MDS can discriminate thermodynamically unstable triplexes

As described in the previous paragraphs, we identified and characterized putative RNA triplexes based on miRNA target prediction, seed site distance discrimination and secondary structure prediction. Furthermore, we proposed a cut-off for the TFE and the seed binding constraint to filter out non-reliable predictions. All these steps could be automatized and we were thereby able to conduct a high-throughput screen for the whole human genome, i.e. for all putative human RNA triplexes. After applying all these steps and selection criteria, there was still a large number of candidate triplexes left. This suggests that the phenomenon of cooperative target regulation is widespread in human cells. However, before a candidate is selected for experimental validation we suggest conducting three more analytical steps, which include MDS, equilibrium concentration prediction and kinetic model simulations. The former is computationally very expensive ($O(n^2)$, where n refers to the number of atoms in the system) and therefore cannot be easily performed in a high-throughput manner. Nevertheless, it provides a more exact prediction of the thermodynamic stability of an RNA triplex and its inherent mRNA-miRNA duplexes. Furthermore, the energy profiles derived from MDS can be used in the parameterization of the kinetic models of cooperative target regulation, as described in Section 4.2.5 on the Research strategy.

For our case study, to demonstrate a proof of principle, we selected a small set of RNA triplexes for MDS. More specifically, for each of the three relevant structural conformations (i) canonical RNA triplex; (ii) RNA triplex with miRNA self-complementarity; and (iii) RNA triplex with target self-complementarity, we nominated those triplexes with the highest and lowest TFE values ($\Delta G_{<min,max>}$) and those with the highest and lowest free energy gain ($\Delta\Delta G_{<min,max>}$). See selected candidate triplexes and results of the MDS production run in Table 4.1.

Selection criteria	mRNA	miRNA$_1$ /miRNA$_2$	Structure pattern	ΔG (kcal/mol)	$\Delta\Delta G$ (kcal/mol)	ST_{duplex} (ps) mRNA::miRNA$_{1/2}$	$ST_{triplex}$ (ps) mRNA::miRNA$_{1/2}$
ΔG_{max}	*RPS6KA5*	miR-410/ miR-590-3p	MSC	-12.263	-4.782	139/145	62/92
	HTRA2	miR-374a/ miR-374a	TSC	-11.363	-3.982	72/72	116/75
	ZNF121	miR-374a/ miR-374a	CT	-10.363	-4.682	250/250	58/91
ΔG_{min}	*EDA2R*	miR-125a-3p/ miR-370	CT	-57.963	-22.282	206/468	218/449
	MUC1	miR-145/ miR-326	TSC	-55.963	-15.282	249/192	179/500
	ABT1	miR-214/ miR-491-5p	MSC	-51.163	-16.182	327/338	239/399
$\Delta\Delta G_{max}$	*GOLM1*	miR-296-3p/ miR-330-5p	MSC	-47.263	-21.482	500/395	118/500
	CCDC3	miR-138/ miR-551b	TSC	-55.263	-21.782	500/479	500/500
	PLXNB1	miR-197/ miR-320d	CT	-49.863	-23.482	333/74	207/387
$\Delta\Delta G_{min}$	*NPHP1*	miR-194/ miR-340	MSC	-16.963	-1.782	113/203	47/44
	KAT2B	miR-106b/ miR-590-3p	TSC	-22.263	-1.682	106/323	500/100
	C19orf69	miR-190/ miR-590-3p	CT	-25.063	-1.582	257/214	74/196

Table 4.1 Results of the MDS production run for exemplary RNA triplexes.
For the three relevant structural conformations (MSC, TSC, CT), triplexes with the most extreme values in terms of TFE (ΔG) and free energy gain ($\Delta\Delta G$) have been selected for MDS. In the last two columns the stability times ($ST_{<duplex,triplex>}$) are presented, i.e. the time duration in picoseconds of the MDS production run in which miRNA$_1$ and miRNA$_2$ are hybridized with the target in a duplex or triplex structure. Shaded cells indicate triplexes that were stable for more than 100 ps and in which the stability time of at least one miRNA exceeds that of its corresponding duplex structure. These triplexes are more likely to show miRNA cooperativity.

For each candidate, based on the secondary structures, we derived 3D models of the RNA triplex and the inherent mRNA-miRNA duplexes. Furthermore, the models were optimized through energy minimization. Both processes were performed as described in section 4.2.3 on the Research strategy. The energy minimized 3D models of RNA triplexes and duplexes were then subjected to MDS experiments in the Accelrys®

Discovery Studio 3.5 simulation software. We defined a threshold value (100 ps) for the simulation time in which the RNA strands in the triplex should maintain hydrogen bonds among miRNAs and the target mRNA. This threshold marks the minimum required simulation time for triplexes to be considered as stable. Furthermore, we compared the stability time (*ST*) between triplexes and their inherent duplex structures. We considered cases in which the triplex is stable for a longer time than any of the inherent duplexes as potential intances of miRNA cooperativity. See both MDS constraints for the identification of functional RNA triplexes below:

(1) $$ST_{triplex} \geq 100ps$$

(2) $$ST_{triplex} > min \begin{cases} ST_{miR1\_duplex} \\ ST_{miR2\_duplex} \end{cases}$$

The ST values for the selected triplexes and their inherent duplexes are given in Table 4.1. The simulation results confirmed our hypothesis that triplexes with low minimum free energy values and strong energy gain are more stable than their inherent duplexes irrespective of the underlying structural conformation (Table 4.1). Hence, we could observe that the predefined MDS constraints were met by all three triplexes with the lowest minimum free energy involving the genes *EDA2R*, *MUC1* and *ABT1*, and those with the strongest energy gain involving the genes *GOLM1*, *CCDC3* and *PLXNB1*. We can conclude that the thermodynamic profile of an RNA triplex is crucial for synergistic target regulation by two cooperating miRNAs. For illustrative purposes we captured simulation videos for two triplexes showing exemplary cases of: (1) a stable RNA triplex (miR-138::CCDC3::miR-551b), and (2) an unstable RNA triplex (miR-374a::HTRA2::miR-374a). The videos can be watched under the following URLs:

(1) www.sbi.uni-rostock.de/triplexrna/files/Supplemental_video_II.mp4

(2) www.sbi.uni-rostock.de/triplexrna/files/Supplemental_video_I.mp4

A proof of principle

As part of the case study on the miRNA target hub *CDKN1A* described in the previous chapter we predicted cooperative *CDKN1A* regulation by pairs of miRNAs. In that study our predictions were based only on steps I and II of the PreCoM workflow. We selected one triplex composed of miR-93, miR-572 and *CDKN1A* and successfully performed validation experiments as described in detail in Chapter 3 and in Lai *et al.* (2012). For a proof of principle, we now conducted secondary structure prediction and MDS for this example. The secondary structure prediction identified a thermodynamically stable structure for this triplex ($\Delta G = -34.46\ kcal/mol$; $\Delta\Delta G = -15.88\ kcal/mol$) with preserved seed bindings (Figure 4.12). The MDS shows that miR-572 and miR-93 stay hybridized to *CDKN1A* for 126 ps and 473 ps, respectively. Both *ST* were significantly longer compared to those simulated for the respective duplex structures (miR-572::CDKN1A, $ST = 42\ ps$; and miR-93::CDKN1A, $ST = 301\ ps$). From this we can conclude that the triplex is more stable compared to its inherent duplex structures, which may be an explanation for the synergistic effect achieved by the pair of miRNAs in the repression of *CDKN1A* as was shown in the wet lab experiments. The simulation videos can be watched at:

Complex	Video URL
miR-93::CDKN1A	www.sbi.uni-rostock.de/triplexrna/files/CDKN1A_miR2.mp4
miR-572::CDKN1A	www.sbi.uni-rostock.de/triplexrna/files/CDKN1A_miR1.mp4
miR-93::CDKN1A::miR-572	www.sbi.uni-rostock.de/triplexrna/files/CDKN1A_triplex.mp4

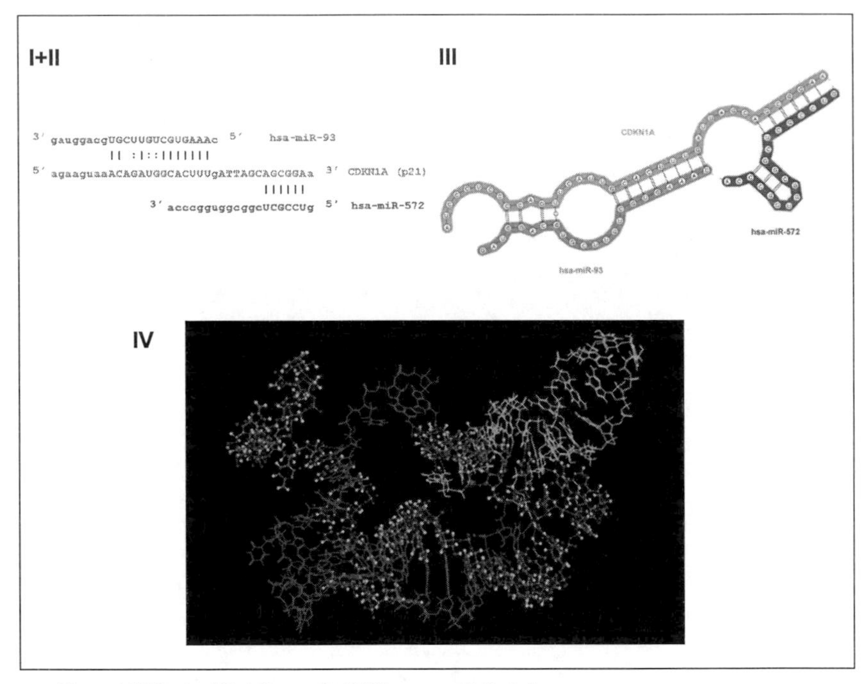

Figure 4.12 First validated case of miRNA cooperativity in human.
We have previously validated the efficacy of the illustrated RNA triplex in Lai *et al.* 2012. (**I+II**) The illustration shows the alignment of *CDKN1A* to the cooperating miRNAs (miR-93 and miR-572). (**III**) Depicted is the minimum free energy structure (2D) of the RNA triplex. (**IV**) Three dimensional model of the RNA triplex. The seed binding of both miRNAs, miR-93 and miR-572, is preserved.

The conclusion to be drawn from this section is that once a RNA triplex or a small set of triplexes is considered for further validation, MDS is a comprehensive tool to differentiate thermodynamically stable from unstable triplexes. It has to be noted, however, that a simulation run for one triplex in an implicit solvent box took ~24h for a 100 ps simulation on a high-performance workstation (12 cores, 2.3GHz, 16GB RAM). Therefore, larger sets of triplexes can only be efficiently simulated in a simplified setting (e.g. in vacuum) and/or in a massively parallelized manner, e.g. on a computer cluster.

4.3.6 Triplex formation depends on equilibrium probability distribution

A pair of miRNAs may exert cooperative target regulation when both miRNA molecules hybridize with one target mRNA (Figure 4.13, bottom-left) because triplex formation leads to a more stable complex (see previous section) which then secures unhampered functioning, i.e. target repression. However, the presence of these three molecules alone doesn't ensure the formation of a triplex structure. Therefore, for all RNA species in a RNA triple (miR_1+miR_2+$mRNA_{tgt}$) and for all possible duplex and triplex structures that can emerge from it we predicted the equilibrium probability distribution in dilute solution (Figure 4.13). To this end, we defined an initial concentration of 100 nM for each RNA species and computed equilibrium concentrations for monomers and complex species ranging between 0-100 nM.

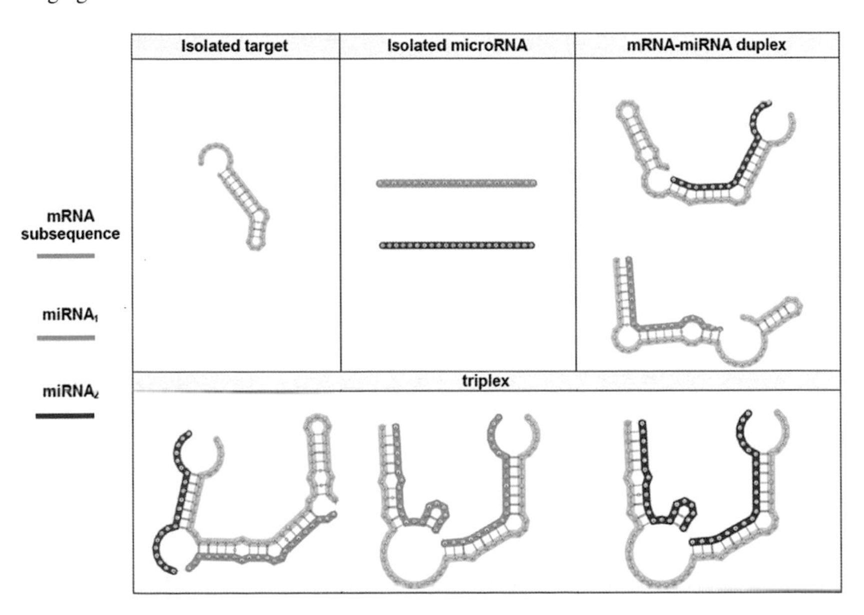

Figure 4.13 RNA monomers and complexes considered for equilibrium concentration predcition.
Equilibrium concentrations where computed for target mRNA and miRNA monomers (top-left and top-middle, respectively), for duplexes composed of one miRNA and the target mRNA (top-right) and triplexes as depicted (bottom).

We consider high triplex equilibrium concentrations, i.e. a high triplex formation probability, beneficial for cooperative target regulation and the realization of an enhanced target repression. Favored RNA triplexes can be disclosed by considering an equilibrium concentration threshold, e.g. $TEC > 50\,nM$. In our case study, when applying this threshold, the number of candidate triplexes dropped to 15% of the initial set (~98,000). Interestingly, we found that the canonical triplexes tend to have higher equilibrium concentrations as compared to other structural conformations (Figure 4.14).

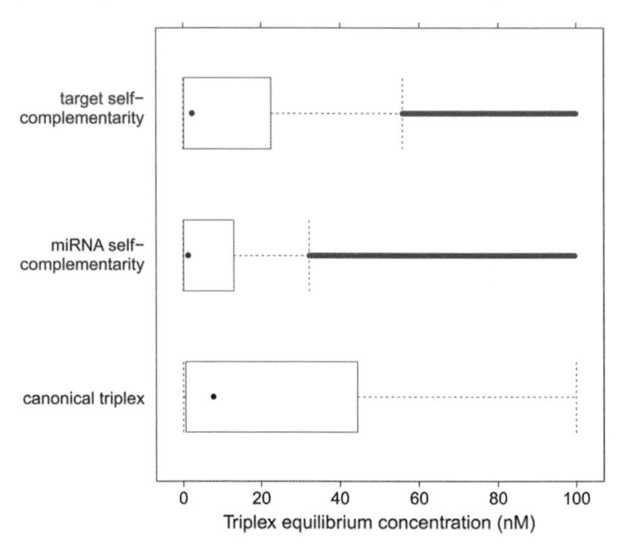

Figure 4.14 Dispersion of equilibrium concentrations for different structural conformations. The box-whiskers plot illustrates differences between the equilibrium concentrations for groups of triplexes with homologous structural conformation. The medians are denoted by the dots inside the boxes while the boundaries of the boxes illustrate the 1st and 3rd quartile of the respective population. The ends of the whiskers represent the lowest and highest data points within the 1.5 interquartile range of the lower and upper quartile, respectively. Outliers are represented by circles (o), however, due to the large number in the first two structure groups, outliers appear as continuous thick line. This plot clearly shows that the class of canonical triplexes has typically higher equilibrium concentrations as compared to the other structural conformations.

The TECs are later used as parameters for the complex association rate constants in the kinetic model of miRNA cooperativity. For the exemplary triplex (miR-93::CDKN1A::miR-572) that was validated as part of the miRNA target hub study we predicted the following concentrations in equilibrium: (i) for the triplex $Trplx = 79\,nM$, (ii) for the duplexes $dplx_1 = 18\,nM$ (miR-93::CDKN1A) and $dplx_2 = 2.1\,nM$ (miR-572::CDKN1A), and (iii) for the monomers $miR_{572} = 13\,nM$, $miR_{93} = 3\,nM$, and $CDKN1A = 0.87\,nM$.

4.3.7 Simulations of miRNA cooperativity reveal landscapes of target repression

In the final step of the PreCoM workflow we performed computer simulations based on a kinetic model of cooperative target regulation. In brief, the model is comprised of ODEs describing the synthesis and degradation of the involved RNA strands as well as their association towards RNA complexes and the dissociation of the same (see Section 4.2.5 for details). We simulated target steady state concentrations for a range of synthesis rate constants of the cooperating miRNA couple applied in small intervals. The simulations give rise to the target repression efficiency achieved by concerted miRNA regulation in comparison to the repression induced by a single miRNA.

First, we performed simulations for the same set of exemplary triplexes as were selected for the MDS, i.e. for each of the three relevant structural conformations (CT, MSC, and TSC) those triplexes with the highest and lowest TFE values ($\Delta G_{<min,max>}$) and those with the highest and lowest free energy gain ($\Delta\Delta G_{<min,max>}$). For the parameterization of the association and dissociation rate constants of the duplexes and the triplex we used their equilibrium concentrations and the PE values respectively, which were computed in the previous steps of the workflow. See section 4.2.4 on the Research strategy for details and Table D.1 in the Appendix for the parameter values of the twelve models. To facilitate efficient visual inspection of the simulation results we generated contour plots as depicted in Figure 4.15 on the next page.

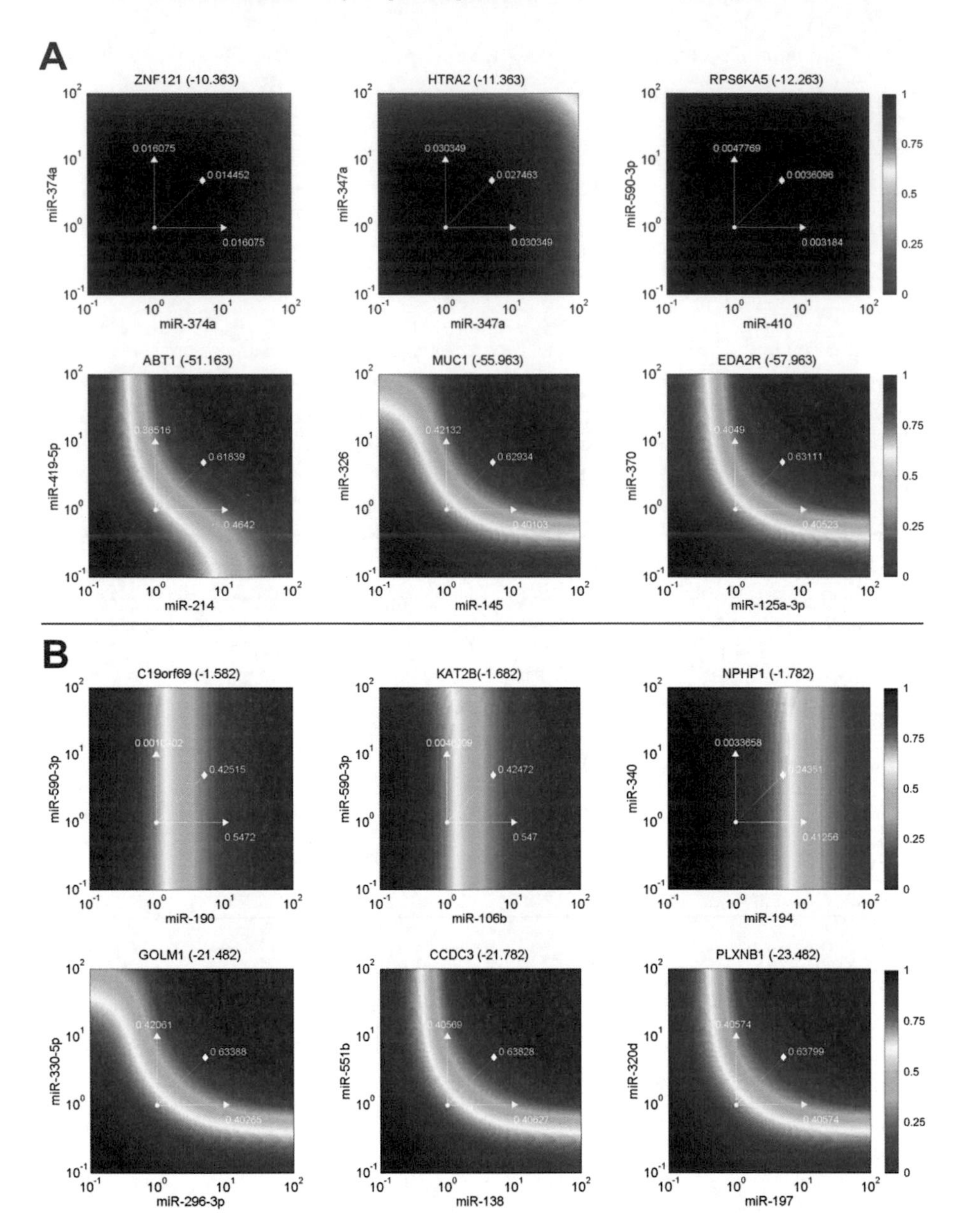

Figure 4.15 Steady states of target candidates for different miRNA synthesis rates.
(**A**) Plots in this panel illustrate the simulation results for candidate triplexes which were selected based on the highest and lowest TFE value (top and bottom row, respectively) in the respective structural conformation class (FLTR: CT, MSC, and TSC). (**B**) Plots in this panel illustrate the

simulation results for candidate triplexes which were selected based on the lowest and highest energy gain value (top and bottom row, respectively) in the respective structural conformation class (FLTR: CT, MSC, and TSC). Plot titles are composed of the targets official gene symbol and the ΔG value (A panels) or $\Delta\Delta G$ value (B panels). Target protein concentrations are represented as colour coded pixels following a rainbow scale as indicated in the legend bar (right), according to which red colour denotes full target expression (no repression) and blue denotes full target silencing (maximum repression). Landmarks within the plot represent three relevant scenarios: (i) normal expression of both miRNAs (1,1); (ii) and (iii) high overexpression of either $miRNA_1$ (10,1) or $miRNA_2$ (1,10); and (iv) moderate up-regulation of both miRNAs (5,5). Additionally, we introduced a measure RG for determining the repression gain achieved in scenarios (ii), (iii) and (iv) compared to the case of nominal miRNA expression (i). RG is computed by subtracting the nominal target steady state from the target steady state after miRNA up-regulation ($RG = SS_{up} - SS_{nominal}$). In case $RG(5,5) > max[RG(10,1); RG(1,10)]$, stronger repression is achieved through synergistic target regulation.

The results of our simulations show different landscapes of target repression which largely depend on the energy parameter values. For example, in the case of high TFE values, only weak or no target repression can be observed even for highly overexpressed miRNA pairs (Figure 4.15A, upper panel), whereas in the case of triplexes with the lowest TFE values in their respective class, strong target repression can be observed, given both miRNAs are moderately up-regulated (Figure 4.15A, lower panel). This indicates a significant cooperative effect in the concerted target regulation and substantiates our assumption that the thermodynamic stability of a RNA triplex determines the target repression that can be achieved. Interesting results were also obtained for the selected triplexes with lowest and highest energy gain in their respective structural classes (Figure 4.15B). For the triplexes that achieve minimal energy gain ($\Delta\Delta G$) compared to their inherent duplex structures, we observed independent target regulation by one of the two miRNAs, while the other had no repression capability at all. In these cases no cooperativity could be observed either (Figure 4.15B, upper panel). In contrast, the triplexes with strongest energy gain in their respective classes (Figure 4.15B, lower panel) exhibited repression landscapes similar to those predicted for triplexes with the lowest free energy values (Figure 4.15B, lower panel). In these cases efficient target repression can be achieved when both miRNAs are moderately up-regulated, which again indicates a considerable synergistic effect achieved in concerted target regulation.

In order to quantify the synergistic effect obtained by cooperative target regulation, we introduced the variable RG, which represents the repression gain that is achieved by overexpressing one or both miRNAs in a RNA triple, and is determined by subtracting the target steady state under normal conditions from that under the influence of a modulated miRNA expression. For our simulations we computed RG for three scenarios: (i) and (ii) strong overexpression of one of the two miRNAs (RG_1 and RG_2), and (iii) moderate up-regulation of both miRNAs (RG_3; Figure 4.15). As expected, the repression gain achieved by combined up-regulation of both miRNAs is most significant for the triplexes with low TFE values (*EDA2R*, *MUC1* and *ABT1*) and strong energy gain (*GOLM1*, *CCDC3* and *PLXNB1*).

In summary, ΔG and $\Delta\Delta G$ values of RNA triplexes are good indicators for the synergistic effect that can be achieved in concerted target regulation through pairs of miRNAs. Simulations based on a kinetic model of cooperative target regulation can provide a detailed picture of the target repression landscape induced by miRNAs pairs. For example, MDS demonstrated a long stability time for the triplex composed of *KATB2*, miR-106b and miR-590-3p. However, simulations of the respective kinetic model show that no synergistic effect can be observed in this case ($RG_3 < RG_2$), which demonstrates the necessity of this last step in the PreCoM workflow.

The parameter values for the dissociation constants of the complexes used in the presented simulations were derived from the PE values determined from the energy optimized 3D models of the RNA triplexes and their inherent duplexes. As mentioned before, 3D model prediction and energy minimization are time consuming processes. Therefore, to be able to perform high-throughput simulations for all predicted triplexes, we instead used the TFE value derived from the secondary structure predictions for parameterizing the complex dissociation constants. The contour plots obtained as a result of our simulations were integrated into our database of human RNA triplexes which is described in detail in the following paragraph. Furthermore, we exported one parameterized mathematical model of miRNA cooperativity into the standardized XML based Systems Biology Markup Language (SBML), which is suitable for the exchange of models using a process description approach (Hucka *et al.*, 2003). This exemplary model was deposited at the BioModels database (Li *et al.*, 2010; ID: BIOMD0000000530).

4.3.8 A database of human RNA triplexes

As described in the previous paragraphs, the PreCoM workflow for the identification of targets of cooperative miRNA regulation and the analysis of these RNA triplexes was implemented and used for a whole human genome analysis. Through the prediction of miRNA target sites, structure and equilibrium concentration prediction, and computer-based simulations of target repression, we generated a detailed overview about the putative RNA triplex inventory in human cells. We deposited the results in a designated web accessible database named TriplexRNA, which provides data on RNA triplex composition, structural and thermodynamic properties and simulated target repression efficiencies. Users can search the database using official gene symbols or miRNA names to receive putative RNA triplexes that involve the requested molecule. Results can be filtered as per predicted TFE or seed site preservation and sorted based on names, seed site distance or TFE values. Each RNA triplex can be examined individually, e.g. for molecule sequences, secondary structures and contour plots of target repression simulations. Moreover, given experimental evidence on miRNA target interactions is referenced. Despite customized download files, the database facilitates programmatic access through a RESTful interface for the extraction of data on molecule sequences, genomic coordinates, TFE values, experimental evidence, secondary structures, and estimated equilibrium concentrations for triplexes and their inherent mRNA-miRNA duplexes. See Figure 4.16 for screenshots of the TriplexRNA database, which can be accessed under the URL www.sbi.uni-rostock.de/triplexrna.

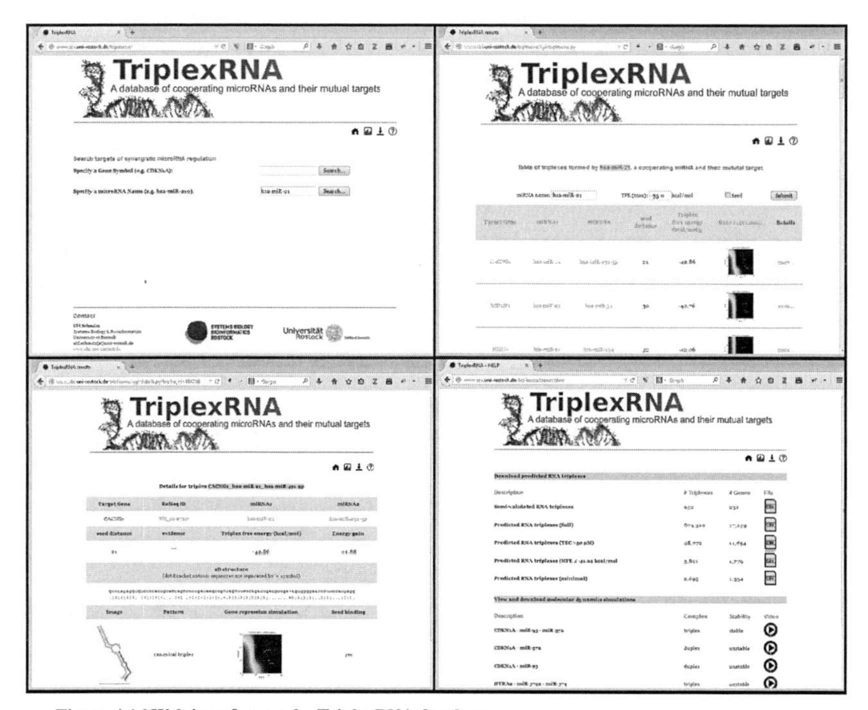

Figure 4.16 Web interface to the TriplexRNA database.
Using the form on the front page (upper left) one can search the database either for human genes or miRNAs to receive information about their possible involvement in RNA triplexes. A preliminary results list will be returned (upper right), which can be expanded or shortened by adjusting the parameters for the triplex free energy cut-off and/or seed binding preservation. Furthermore, users can sort the results in a parameter dependent ascending or descending order by clicking the column headings. More information on a specific triplex can be retrieved by accessing the 'Details' page associated to each RNA triplex (lower left). Filtered and non-filtered data in the database can be downloaded in CSV file format on the 'Downloads' page (lower right). This page also provides five simulation videos of stable and unstable complexes.

4.4 Synthesis of results

The PreCoM workflow described in this chapter is intended for a systematic large-scale identification of targets of cooperative miRNA regulation (Figure 4.2). It integrates several computational approaches: (i) RNA hybridization prediction, (ii) RNA structure modelling, (iii) molecular dynamics simulations, and (iv) kinetic modelling.

We hypothesize that cooperative target regulation is realized through the formation of RNA triplexes composed of the target mRNA and two cooperating miRNAs (Figure 4.3), which stabilize weak miRNA target hybrids and thereby ensure functioning in terms of efficient target repression. To elucidate RNA triplex efficacy our workflow contains a number of interconnected analytical steps that can be used to narrow down a set of high-confidence candidate triplexes that may be considered for subsequent experimental investigations. We demonstrated the applicability of the PreCoM workflow by implementing all steps for the identification and analysis of human RNA triplexes. Results of our analysis indicate that cooperative target regulation is a wide-spread phenomenon in human cells and that it may affect thousands of target genes. Furthermore, we were able to point out some key features that are required for a functional RNA triplex: (i) a distance of 13-35 nt between the seed sites of putatively cooperating miRNAs in the 3′ UTR of their mutual target mRNA; (ii) a stable triplex structure with preserved seed bindings of the involved miRNAs; (iii) a strong binding affinity among the involved RNA molecules; and (iv) a strong repression gain achieved by synergistically acting miRNAs.

4.4.1 Ways to narrow down high-confidence RNA triplexes

After applying the first two steps of the PreCoM workflow, (i) miRNA target predictions, followed by (ii) seed site distance filtering, we identified 15,062 targets of synergistic miRNA regulation in humans. We found that miRNA pairs that cooperate often share several targets. The highest number of mutual targets ($n = 14$) is shared by hsa-let-7g and hsa-miR-376c and includes the genes *AMMECR1L, CYP26A1, ZFYVE26, MUC7, FUS, TMEM38B, CWH43, CWH43, LIN28B, ZNF823, DVL3, SRGN, INPP5A,* and *HDAC8.* However, we hypothesized that not every predicted candidate triplex is necessarily functional in terms of cooperative target regulation. To address this uncertainty we applied the remaining steps of the workflow and thereby filtered out a set of high-confidence triplexes, more likely to be functional. See Figure 4.17 (next page) for an illustration on how the number of cooperation partners per miRNA drops when selection criteria are applied.

One approach to narrow down high-confidence triplexes is already applied in the first step of the PreCoM workflow – the target prediction. We used the miRanda algorithm (Enright *et al.*, 2003) which is a sensitive method, i.e. it predicts many possible miRNA target interactions, covering on the one hand most true positive cases but on the other hand accepting an increased number of false-positive predictions (Sethupathy *et al.* 2006b; Yan *et al.* 2007). In our analysis we already filtered predicted miRNA-target interactions by considering only miRNAs that are conserved across species, target sites with good prediction scores ($S_{miRSVR} \leq -0.1$; Betel *et al.* 2010), and by excluding predicted target sites in the 5′ UTR and the coding sequence of target mRNAs. The latter was applied because experimental validation for functional RNA triplexes has so far only been provided for miRNA target sites residing in the 3′ UTR of the target mRNA. One way to identify a more reliable set of predicted miRNA target interactions is to take several algorithms which are based on different approaches and to consider only the intersection of their predictions. Another possibility is to consider only experimentally supported miRNA-target interactions for further investigation. The TriplexRNA database contains 952 triplexes for which inherent duplexes involving 252 human genes have experimental support as is indicated in the miRTarBase database (release 4.5; Hsu *et al.* 2011). In this subset, 346 miRNA binding sites were identified by CLASH, a recently developed experimental technique for the high-throughput identification of exact miRNA binding sites (Kudla *et al.*, 2011; Helwak *et al.*, 2013).

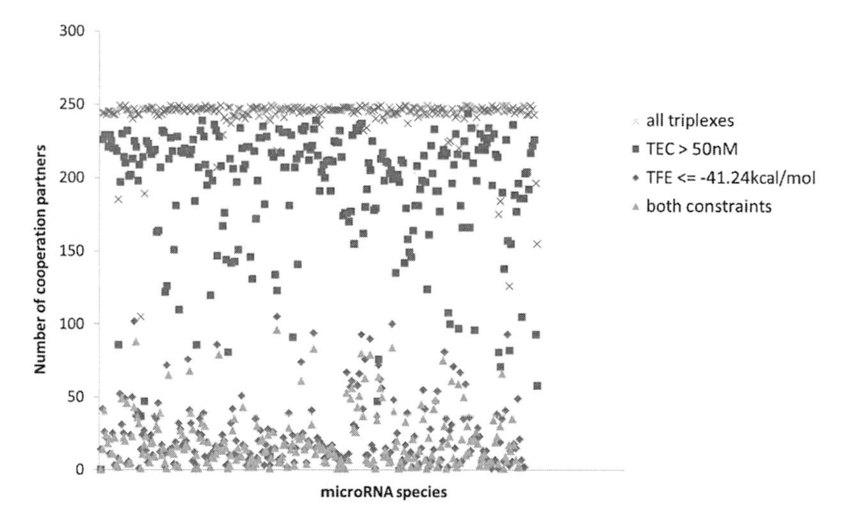

Figure 4.17 Number of cooperation partners per miRNA.
In our analysis we considered 250 conserved miRNAs for the prediction of RNA triplexes including two cooperating miRNAs and a mutual target mRNA. After the first two steps in the PreCoM workflow (miRNA target prediction and seed site distance filtering) we found that miRNAs are potentially cooperating with most of the other miRNAs in the regulation of one or more mutual target genes (crosses). After applying filtering criteria the number of cooperation partners drops as indicated by the squares for triplexes with $TEC > 50\,nM$, diamonds for triplexes with $TFE < -41.24\ kcal/mol$, and triangles for triplexes that fulfil both constraints.

Along with the prediction of triplex secondary structures, one can compute the free energy (TFE). By applying a cutoff, e.g. $\Delta G_{triplex} = -41.24\ kcal/mol$ as in our case study, which is three standard deviations below the mean TFE ($Z = 3$), one can reduce the set of candidate RNA triplexes to those which are thermodynamically more stable than others. The formation of RNA triplex structures necessitates a certain binding affinity among the RNA molecules. The prediction of triplex equilibrium concentrations helps identify RNA triples with strong hybridization tendencies. In our case study we applied a threshold of $TEC > 50\,nM$ for considering a triplex as having the required binding affinity. See in the Table 4.2 how the number of candidate triplexes and target genes reduce when applying different filtering criteria.

Description	# Triplexes	# Genes
Predicted RNA triplexes (full)	674,310	17,259
$TEC > 50\,nM$	98,073	11,654
$MFE \leq -41.24\,kcal/mol$	3,821	1,779
Semi-validated RNA triplexes	952	252

Table 4.2 Crystalizing high-confidence triplexes by applying different filtering criteria.

4.4.2 What can be learned from 2D and 3D models of RNA triplex structures?

Most of the classical miRNA-target prediction algorithms use sequence complementarity (with focus on the seed site) and secondary structure prediction as two elementary steps in their workflows. We make use of these predictions in the first step of the PreCoM workflow. After we filtered the results for cases of miRNA pairs with neighbouring binding sites in a mutual target, we again performed secondary structure prediction in the third step of our workflow, this time for the putative RNA triplex. Some miRNA target prediction algorithms include means to estimate the target site accessibility. This involves the computation of full length target 3′ UTR folding (Robins *et al.* 2005; Hammell *et al.* 2008) or even predicting the secondary structure for the entire mRNA target sequence (Thadani and Tammi, 2006). Given the target secondary structure, these algorithms determine the energetically favourable arrangement with or without miRNA hybridization by comparing the energy consumed for unfolding (making the target site accessible) and the energy gained by miRNA/mRNA hybridization ($\Delta\Delta G$). Unfortunately, the prediction accuracy decreases with the length of the RNA strand that is subjected to secondary structure prediction (Wang and El Naqa, 2007). Therefore, we decided not to include target site accessibility in our analysis.

We utilize local secondary structe predictions as a means for the high-throughput (whole genome) identification of RNA triples that are in general able to form RNA triplexes. From the NUPACK package we used the tools *mfe* and *complexes* for the prediction of triplex local secondary structures and TFEs, respectively. NUPACK is a software suite that is dedicated to the analysis and the design of RNA complexes (Zadeh *et al.*, 2011). Pseudoknot structures were not considered in the analysis because their prediction is described as a NP-hard problem (Lyngsø and Pedersen, 2000). Due to 2D structure predictions, we get instantaneous insight into seed binding preservation after folding. Furthermore, to identify triplexes that are more likely to be functional we considered the computed binding free energy (TFE) as an important criterion. Conventional miRNA-target prediction algorithms use the same criterion to discriminate functional from non-functional structures (Fan and Kurgan, 2014). Furthermore, the 2D triplex structure predictions are crucial in the workflow because they give rise to possible interaction sites which are used for building the 3D model.

3D models obviously provide a more realistic representation of RNA triplexes. We therefore predicted RNA triplex 3D structures based on the previously identified secondary structures. This is common practice and to date no method exists which could predict an RNA complex 3D structure *ab initio*, i.e. from the primary sequences. Frequently cited RNA 3D structure modelling tools (such as RNAcomposer, MC-SYM, etc.) use single stranded RNA sequences and secondary structure folding information in dot bracket notation for their predictions. MC-SYM for example, to model the RNA duplexes, artificially introduces a GAAA tetraloop between two strands of RNA and thus models the 3D structure of large single stranded RNA (Parisien and Major, 2008). Thereafter, the tetraloop is removed again to separate the two strands. We also modelled mRNA-miRNA triplexes as a single strand using RNAcomposer (Popenda *et al.*, 2012) and then separated mRNA and two miRNAs strands by manually deleting the bonds. This approach is the same as in (Parisien and Major, 2008) with the exception that we did not introduce any artefact by including an extra nucleic acid sequence. The strategy was adopted to ensure a consensus between 3D structures generated by RNAcomposer and the secondary structure pattern predicted in the previous step of the PreCoM workflow. Structure editing steps in our presented workflow are necessary to separate the three RNA strands. Likewise, the removal of GAAA tetraloops in MC-SYM entails these editing steps. To further optimize the geometry of the RNA triplexes, we used the energy optimization protocol available in Accelrys® Discovery Studio 3.5 to remove any steric overlap that produces bad contacts. In summary, we generated a more realistic model of mRNA-miRNA interaction by only considering hydrogen bonds between the two RNA species and not by introducing an artificial intermediate nucleotide sequence for structural analysis and molecular dynamics simulation studies. The same strategy was reported in other studies (Paciello *et al.*, 2011; Gan and Gunsalus, 2013).

The workflow that was conducted for 3D structure prediction is illustrated in Figure 4.4. Despite its role as realistic representation of RNA triplexes, the 3D model was used as a basis for MDS. Furthermore, we used predicted potential energies of RNA duplex and RNA triplex structures for characterizing the respective complex dissociation constants in our kinetic models of cooperative target regulation.

4.4.3 Time consuming but valuable MDS

MDS add another dimension to the 3D RNA triplex models, namely time. We used MDS to estimate whether RNA triplex structures are stable over time under the influence of atomic movements. Even more realistic simulations are achieved by expanding the time duration and by including a solvent to the models of RNA triplexes to observe possible effects on the structure, energetics and dynamics of the involved biomolecules. A model which is closer to the native miRNA-induced target regulation would be achieved by including RISC or critical components of the complex like the Argonaute protein (see an exemplary model in Appendix D.1). However, with each detail that is added to the model or the simulations, the computational cost increases significantly. For the presented case study in which we predicted and analysed targets of cooperative miRNA regulation we performed steps 1-3 and 5-6 of the PreCoM workflow as they could be computed in a reasonable amount of time. Here, the most time consuming tasks were the secondary structure predictions and the simulations based on kinetic models. Both tasks required

~21d for 674,310 putative RNA triplexes on a workstation with two QuadCore CPUs (8 cores, 2.4 GHz) and 16 GB of memory.

Unfortunately, the MDS can be performed for a small number of triplexes only. We performed 3D model predictions and MDS for twelve selected cases (see section 4.3.5), where each run requires ~24h time. In the MDS production run we defined a threshold of 100 ps. Only those RNA complexes that were stable until this time were further simulated for 400 ps. Stability is given as long as one hydrogen bond is maintained between mRNA and miRNA strands. We adopted this strategy to determine the stability of the nucleic acids complexes as it has been previously successfully used in other studies (McDowell *et al.* 2007; Shen *et al.* 2011). We have observed that RNA triplexes with low TFE and strong energy gain are stable for a longer time duration in the MDS production run.

4.4.4 Target regulation efficiency revisited

We utilized data from the initial steps in the PreCoM workflow for the parameterization of a kinetic model that we used to predict the efficacy and efficiency of cooperative target regulation by pairs of miRNAs. Our model is composed of ODEs that describe the dynamics of monomers and complex species involved in synergistic target regulation. We chose an ODE-based approach as it is the most useful and widely established tool for the mathematical description of mechanisms underlying biochemical reaction systems. Several mathematical models have been proposed for miRNA-target regulation (Levine *et al.* 2007; Nissan and Parker 2008; and Zinovyev *et al.* 2013). However, these models deal with paired miRNA target regulation only (one miRNA, one target). We developed a model for collective and cooperative target regulation by several or even many miRNAs in our previous work (Lai *et al.*, 2012; Chapter 3). This model was well parameterized with experimental data in which single miRNA species were transfected to cells and the change in target protein concentration was measured using western blotting experiments. However, we aimed to generate models of individual triplexes in a high-throughput manner. We therefore utilized the data on the thermodynamic stability of complex species (steps 3 and 4) and the data on RNA binding probabilities derived from equilibrium concentration predictions (step 5). We used the predicted equilibrium concentrations for parameterizing the association rate constants in our models and the TFE values for parameterizing the complex dissociation constants. With the parameterized models we were able to simulate target steady states for a range of miRNA synthesis rates which we visualized in form of contour plots (Figure 4.15). It has to be noted that for selected RNA triplex candidates we used the PE values derived from the MDS for defining the dissociation rate constants, while for all other candidates, in order to facilitate high-throughput model simulations, we used the TFE values derived from the secondary structure prediction (step 3), which is much faster than the MDS.

4.4.5 Cooperative target regulation - a phenomenon with relevance for cancer?

As a result of the study we found evidence for cooperative target regulation being a wide-spread phenomenon affecting a large number of human genes. However, we wondered whether these genes are associated with particular cell biological processes. To this end, we performed a functional enrichment analysis with focus on KEGG pathways using the Database for Annotation, Visualization and Integrated Discovery (DAVID; Huang *et al.*, 2009). We found that the 252 genes involved in triplexes with experimental support are predominantly associated with different cancers, including prostate cancer, colorectal cancer, small cell lung cancer, bladder cancer etc., as well as cancer-relevant pathways. This observation may result from the fact that many miRNA-target interactions were validated as part of a cancer-related study. Consequently, we repeated the analysis for a set of predicted high-confidence triplexes ($TFE \leq -41.24\, kcal/mol; TEC > 50\, nM$). Surprisingly, the involved genes are also associated with many cancer types as well as cancer-related processes (e.g., regulation of actin cytoskeleton, focal adhesion, and MAPK signalling). A better understanding of this interesting observation certainly requires further *in silico* and wet lab investigations. Data generated in the pathway enrichment analysis (including p-values and fold enrichments) are included in Table D.2 and Table D.3.

4.4.6 General remarks

The PreCoM workflow described in this chapter integrates a number of computational interdependent steps for identifying targets of cooperative regulation by pairs of miRNAs. It assembles established computational methods that have already been successfully implemented for the prediction of RNA duplexes composed of one miRNA and its target mRNA and adds analytical tools that ensure the best possible *in silico* prediction of functional RNA triplexes with high-confidence. Among the identified structural conformations, the canonical triplex and the triplex containing target self-complementarity are typically more stable than other structural conformations and have stronger binding affinities. Especially the canonical triplex forms the highest fraction (68%) in the set of high-confidence triplexes ($TFE \leq -41.24\, kcal/mol$; $TEC > 50\, nM$) followed by triplexes with target self-complementarity (18.8%). This suggests that in these triplexes, target expression is more efficienty regulated, i.e. stronger target repression will be achieved, which we substantiated with our model simulations (Figure 4.15).

Based on our analysis, we can assume that cooperative target regulation by pairs of miRNAs is an integral part of post-transcriptional gene regulation. Our simulations show that cooperating miRNAs can fine-tune target expression or enhance target repression. They thereby support the realization of an adapted target expression profile to meet the requirements given in different cell types and conditions. Therefore, we consider cooperative target regulation to be an important cellular tool which compromises between mild target repression mediated by single miRNA-target interactions (Selbach *et al.*, 2008) and target silencing often only achieved through concerted target regulation by many miRNAs (see previous chapter on miRNA target hub genes; Doench and Sharp 2004; Balaga *et al.*, 2012).

Ultimately, the PreCoM workflow is a comprehensive approach for the *de novo* identification and characterization of RNA triplexes that can be seamlessly applied in other eukaryotic species as well. It has to be noted, however, that more experimental evidence is necessary to adjust this workflow in order to produce the most accurate predictions. Unfortunately, due to the lack of experimental data on this phenomenon, we couldn't estimate the performance of our approach in terms of sensitivity, specificity and accuracy.

5 Discussion

Synopsis

In this chapter all results are summarized and set into a broader context. In particular, I discuss the role of concerted and cooperative gene regulation at the post-transcriptional level and associated implications for the analysis of cell biological phenomena as well as the emergence and progression of diseases. It is explained why such analyses require sophisticated workflows based on an integrated bioinformatics and systems biology approach. One goal of this work was to ensure reproducibility of the workflows and the results. This, and ways for implementing the workflows in other studies are covered in this chapter. Finally, I discuss limitations and directions for enhancement of the proposed workflows.

5.1 Exploration of a new class of ncRNAs

For a long time those parts of the DNA that do not encode proteins were believed to be non-functional and therefore they were often referred to as 'junk DNA'. Today we know that the human genome consists of 20,389 genes (plus 14,345 pseudogenes[14]) in contrast to 24,126 non-coding genes (including 9,656 small non-coding genes and 14,470 long non-coding genes) according to the latest human genome assembly GRCh38 (Genome Reference Consortium Human Build 38). Transcripts of these non-coding genes are categorized into several ncRNA classes with various functionalities that include transporters (e.g. tRNA), enzymes (ribozymes; Gilbert, 1986) and post-transcriptional regulators (e.g., miRNAs; He and Hannon, 2004). Other parts of non-coding DNA also serve specific purposes. These regions are comprised of *cis*-regulatory elements, pseudogenes, repeat sequences, transposons, viral elements and telomeres. Compelling evidence has accumulated in the last decade which suggests that more parts of the DNA than previously assumed serve a purpose and that there may be no such thing as 'junk DNA'. However, the purpose of many DNA subsequences is still unknown and therefore requires further investigation. Similarly, the complex network of gene regulation organized in three control layers (transcriptional, post-transcriptional and post-translational regulation) requires more research, the development of computational tools, and the design of integrated workflows, towards a better understanding of e.g. disease regulatory mechanisms, and for the design of more effective therapeutic approaches.

In this thesis these complex regulatory mechanisms were elucidated in parts, with a special focus on a class of ncRNAs, the post-transcriptional gene regulators named miRNAs. MiRNAs and their regulatory capacity have drawn much attention over the last decade because of their abundance, omnipresence, far-reaching regulatory control, involvement in diseases and more (Tétreault and De Guire, 2013). One interesting fact about miRNAs is that they can inhibit the translation or induce degradation of dozens to hundreds of mRNA targets per miRNA species (Selbach *et al.*, 2008). Therefore it is no surprise that genes on the other hand can be under the regulatory control of several miRNAs at the post-transcriptional level (Borgdorff *et al.*, 2010; Wu *et al.*, 2010). We have elucidated concerted miRNA-induced target hub gene regulation, a phenomenon that has received little attention so far, and proposed for this purpose an integrative approach, combining systems biology and bioinformatics methods. The result is the CATH workflow in which (i) data from various resources are integrated for setting up the structure of a miRNA-involved regulatory network; (ii) confidence scores are assigned to the interactions identified in the network; (iii) a network analysis is performed to characterize network properties and to identify network motifs; (iv) a mathematical model is derived that is used for generating hypotheses about the network and perform predictive simulations. We applied this workflow to study the regulation of the miRNA

[14] Pseudogenes are formerly functional protein coding genes that have been disabled due to genetic mutations.

target hub and cell cycle regulator *CDKN1A* (a.k.a. p21). We found that the regulatory network comprised of TFs, miRNAs and protein-interaction partners is enriched in FFLs and that a distinctive expression pattern of cooperating and non-cooperating miRNAs is used to fine-tune transient and long-term regulation of *CDKN1A* as per the requirement in different cellular conditions (Figure 3.7). Our model was able to successfully predict CDKN1A protein levels for nine different cellular functions (Lai *et al.*, 2012). Furthermore, we used the kinetic model to study the effect of transient upstream signals triggering target hub gene expression in the context of combinatorial and cooperative regulation by miRNAs. The validity of the network and the kinetic model was proven by the described wet lab experiments in which we validated cooperative *CDKN1A* regulation by miRNA-93 and miR-572. Furthermore, for the predictions of CDKN1A expression in different cell biological processes we found supportive evidence in the literature.

The ability of miRNA pairs to cooperate in the regulation of a mutual target was demonstrated in (Saetrom *et al.*, 2007) where the investigators identified a distance range between target seed sites that is crucial for miRNA cooperativity. For the *in silico* prediction of targets cooperatively regulated by miRNA pairs we proposed a sophisticated workflow, named PreCoM, integrating methods from bioinformatics, structural biology and systems biology. In particular, the workflow is composed of six consecutive and interdependent steps: (i) miRNA target site prediction; (ii) identification of neighbouring target sites; (ii) secondary structure prediction; (iv) molecular dynamics simulations; (v) prediction of equilibrium concentrations; and (vi) simulation based on a kinetic model of cooperative target repression. We applied the PreCoM workflow in the human genome and found that according to our results cooperative target regulation is a frequently occurring phenomenon in human cells that may affect thousands of target genes. Furthermore, we identified key features given in most functional RNA triplexes which include a strong binding affinity between the miRNA pair and the target mRNA and a stable triplex structure with preserved seed bindings.

5.2 The necessity of an integrative bioinformatics and systems biology approach

Apart from large scale interaction networks, it is often desirable to study a gene-specific regulatory network for better understanding mechanisms of individual gene regulation. This is especially interesting in the case of hub genes, i.e. those interacting with many other molecules. A gene-specific regulatory network was, for example, analysed by Sinha *et al.* (2008) who suggested that miRNAs involved in TP53 (p53) regulation are important for controlling signalling pathways relevant in tumour progression (Sinha *et al.*, 2008).

To analyse the complex regulatory network of a miRNA target hub we had to design and implement a novel, integrative approach combining bioinformatics and systems biology methods. Bioinformatics algorithms, for example, helped to identify putatively cooperating miRNA pairs that share a mutual target (in our case *CDKN1A*). Mathematical modelling, on the other hand, was crucial for obtaining new insights into the complex mechanisms by which target hub gene regulation is realized in a setting with multiple miRNAs and other trans-acting elements.

The CATH workflow enabled us to test hypotheses about different modes of collective target regulation (independent, interdependent and synergistic) and the regulation of *CDKN1A* in several biological processes and different tissues. The predictive model that is described as part of Chapter 3 complements previously published models used for elucidating miRNA-mediated repression of target genes (Nissan and Parker, 2008; Zinovyev *et al.*, 2010). It explains mechanisms that facilitate fine-tuning of miRNA-induced repression via collective and cooperative target regulation. Moreover, our analysis provides a feasible explanation for the mild *in vivo* repression induced by single miRNAs at basal levels (Selbach *et al.*, 2008) but also for the counterpart, namely the ability of miRNAs, via cooperation, to strongly repress target expression or fine-tune their transient and long-term responsiveness in different cellular scenarios. The work presented in Chapter 3 provides an illustrative example of the necessity of integrating different bioinformatics and systems biology approaches in order to handle and asses the inherent complexity that is associated with miRNA regulation and large-scale gene regulatory networks. MiRNAs in particular have to be considered as part of complex multi-level biochemical networks, which involve transcriptional, post-transcriptional and signalling regulation (Vera *et al.*, 2013b). Furthermore, these networks are typically enriched in non-linear motifs, like FBLs and FFLs. Mathematical modelling is required in this case because the dynamics emerging from these motifs evade common rational thinking. What is more, rather than a loose assembly of methods, the elucidation of functional mechanisms in these regulatory networks requires the design of sophisticated interlinked workflows involving approaches across disciplines.

The integrative workflow for the prediction and analysis of targets of cooperative miRNA regulation (PreCoM) described in Chapter 4 provides an enhancement over established methods for miRNA target prediction which still produce a large number of false positive predictions and are not able to reliably predict the target repression efficiency (Ritchie *et al.*, 2013). With the aim to identify and analyse cooperatively acting miRNAs and their mutual target genes we tried to address the shortcomings of existing miRNA-target prediction algorithms right from the beginning by integrating additional computational methods and expanding the analytical thouroughness for achieving more detailed and reliable results. For example, based on thermodynamic properties we identified favourable structural conformations in cooperative target regulation. Additionally, we used detailed molecular dynamics simulations in the 3D space to gauge the thermodynamic profile and triplex stability over time. We considered new aspects that may feature effective target regulation by predicting the equilibrium probability distribution of RNA complex species. Moreoever, we used a kinetic modelling approach to derive a broad overview about the target repression capacity associated with a pair of cooperating miRNAs.

We can conclude that the complexity of gene regulation necessitates the development of integrative workflows, such as the ones presented in this work. In particular, the CATH workflow can be adopted for studying gene regulatory networks in general and for elucidating the regulation of miRNA target hubs in particular. The PreCoM workflow was designed to identify and analyse instances of a particular gene regulatory mechanism at the post-transcriptional level, but could also inspire similar approaches for the investigation of other molecular biological phenoma.

5.3 Outlook

Despite the conclusive results of our analyses, there are some aspects which were not taken into account in the presented work. In most cases we had to compromise with respect to the availability of data or the expenditure of time. A number of future tasks can therefore be considered for achieving even more realistic predictions. For example, in our miRNA target hub model we have not considered the kinetics and rate limiting steps involved in the miRNA biogenesis pathway which may also have an impact on the target repression efficiency (Winter *et al.*, 2009; Saj and Lai, 2011). Another aspect is the effect of mutations in miRNA genes or transcripts (e.g. caused by RNA editing). Furthermore, somatic and epigenetic modifications may alter miRNA expression, processing, maturation or functioning.

Moreover, for the case study described in Chapter 3, it would be desirable to further characterize (i) individual synthesis and degradation rates of *CDKN1A* targeting miRNAs; (ii) the availability of RISC components like AGO; (iii) miRNA binding sites in other targets competing for miRNA hybridization; and (iv) spacial aspects relevant for target repression efficacy. For all these aspects data, in the public domain, is sparse and would therefore require the generation of new experimental data.

A bottle neck in the PreCoM workflow for the identification and characterization of RNA triplexes are the MDS. These simulations, with a size of ~60.000 particles are very expensive in terms of CPU time ($O(n^2)$). Moreover, the size of the timesteps (1 femtosecond) and overall simulation time (up to 500 ps) cause simulations for a single triplex to last for ~24h. We therefore conducted MDS for a selected set of candidate triplexes only. For the future, it could be desirable to realize a high-throughput solution with a reasonable time resolution. More time-efficient simulations may be achieved through: (1) massive parallelization (e.g. on a computer cluster); (2) increasing the size of simulation time steps (e.g. 2 femtoseconds); and (3) using a vacuum instead of simulations in a solvent box. These approaches will be implemented in future related works.

Some of the predictions made in our miRNA target hub analysis were validated experimentally by our collaborators or we found confirming evidence in the literature (see Chapter 3 and Appendix C.3 for details). In contrast, the PreCoM workflow for the prediction of human RNA triplexes and their *in silico* functional assessment was designed based on previous observations (Lai *et al.*, 2012; Saetrom *et al.*, 2007; Vella, 2004). Nevertheless, the generation of experimental support for the new predictions remains a necessary task for the future. There currently exists no experimental approach that could assess the validity of all predictions in a high-throughput manner. Therefore, we were not able to estimate the performance of our approach in terms of sensitivity, specificity and accuracy. The experimental validation of RNA triplex structures would be achieved best through X-ray crystallography or nuclear magnetic resonance spectroscopy (NMR). However, as a proof of principle we realized, based on data of the crystal structure of Argonaute in complex with miR-20a (Elkayam *et al.*, 2012), a detailed 3D model of native miRNA target regulation. We expanded this model to illustrate the possibility of triplex formation by two miR-20a molecules and the *NUERL1B* target mRNA, which contains two neighboring binding sites for miR-20a (a detailed description of this model and the MDS performed is provided in Appendix D.1). We consider this a noteworthy *in silico* validation for the formation of a RISC/miRNA/target complex, which is necessary for cooperative target regulation. The strength of miRNA cooperativity can be determined experimentally using a similar protocol as in Chapter 3 (see also Appendix C.3). However, at the moment no such data is available in the public domain.

5.4 Conclusion

It is established knowledge that miRNAs are important post-transcriptional regulators of gene expression and that they are involved in many cellular processes and in the emergence and progression of human diseases such as cancer (Schmitz *et al.*, 2013).

The workflows designed in this work can be applied in scenarios where the role of miRNA regulation is investigated in the control of single genes, (signalling) pathways and larger networks driving cellular processes. They can also be seamlessly applied for studies in other eukaryotic species. Furthermore, our results give rise to features and dynamics induced by a new mechanism of concerted target regulation by miRNAs. This broadens our view on miRNA function and impact on cellular processes but can also lead to new hypotheses about the link of miRNA regulation to the emergence and progression of diseases as well as to the design of new therapeutic approaches. The methodologies used in this work are explained in such a way that they can be reproduced by others. Data generated in this work have been made available to the public in a web accessible database. Models have been encoded in standardized formats and deposited in public repositories.

Ultimately, the workflows presented in this thesis along with the approaches and tools utilized in this work form a solid basis for further investigations in the large domain of computational RNA biology. The proposed workflows are in particular comprehensive means for the elucidation and characterization of miRNA involved regulatory networks as well as for the *de novo* identification and characterization of RNA triplexes.

Appendix

A Author contributions to the two key publications

First key publication:

X Lai*, **U Schmitz***, S Gupta, A Bhattacharya, M Kunz, O Wolkenhauer, J Vera (2012) Computational analysis of target hub gene repression regulated by multiple and cooperative miRNAs. *Nucleic Acid Research*, 40: 8818-8834, (* eq. contr. authors).

From the original publication:

The original idea was developed by Julio Vera., Ulf Schmitz and Xin Lai and the mathematical model was derived by Julio Vera and Xin Lai. Numerical simulations and figures were implemented by Xin Lai under the supervision of Julio Vera and Olaf Wolkenhauer. Bioinformatics analyses were performed by Shailendra K. Gupta, Ulf Schmitz and Xin Lai under the supervision of Olaf Wolkenhauer. The experiments were carried out by Animesh Bhattacharya and Manfred Kunz. Biological interpretation of the results was done by Julio Vera and Manfred Kunz. All the authors drafted the manuscript.

Details on the contributions from Ulf Schmitz:

General:

- Essential contributions to the development of the method.
- Cross-evaluation of the contributions from the co-authors.
- Leading role in all discussions.
- Contribution of key elements to the manuscript and detailed revision of other parts.

Bioinformatics analyses:

- Integration of heterogeneous data from publicly accessible databases.
- Prediction of miRNA binding sites.
- Construction of molecular interaction network in Cytoscape .
  - Inclusion of SBGN and SBML standards.
- Network analysis
  - Extraction and analysis of topological properties of the network.
  - Determination of network structure significance.
- Confidence assessment of molecular interactions
  - Development of a scoring system for PPI, TF-Gene, TF-miRNA, miRNA-target interactions.
  - Conducting of a survey among domain experts, including survey design and analysis.
- GO analysis
  - Determination of miRNA activity profiles for various biological processes.

Second key publication:

U Schmitz, X Lai, F Winter, O Wolkenhauer, J Vera, S Gupta (2014) Cooperative gene regulation by microRNA pairs and their identification using a computational workflow. *Nucleic Acid Research*, 42, 12, p. 7539-7552.

From the original publication:

Julio Vera, Shailendra K Gupta, Ulf Schmitz and Xin Lai developed the idea and designed the workflow. Ulf Schmitz carried out the data retrieval, prediction and characterization of RNA triplexes, as well as the statistical analyses. Furthermore, he has designed the TriplexRNA database and web interface. Shailendra K.Gupta performed and analyzed the molecular dynamics simulations of cooperating triplexes. Felix Winter, Julio Vera and Xin Lai derived the kinetic model and performed the predictive simulations. Julio Vera and Olaf Wolkenhauer helped with the analysis and interpretation of the data. All authors contributed in drafting the manuscript, revising it critically and approved the final version. We would like to thank S. Smita, K.P. Singh and M. Karmakar for their support in modeling the 3D structures of RNA complexes. Julio Vera and Shailendra K. Gupta are joint senior authors on this work.

Details on the contributions from Ulf Schmitz:

General:

- Essential contribution to the development of the method.
- Cross-evaluation of the contributions from the co-authors.
- Leading role in all discussions.
- Writing of the manuscript plus coordination and detailed revision of contributions from the co-authors.

Bioinformatics analyses:

- Prediction of whole human genome miRNA-target interactions.
- Secondary structure prediction and structure analysis.
- Prediction of molecule/complex equilibrium concentrations.
- Statistical analysis of structural and sequence features in R.
- Automation of the core workflow.
- Identification of enriched GO term associations in significant RNA triplexes.
- Design and implementation of the TriplexRNA database.
- Design of the web interface and implementation of the programmable interface using REST.

B MiRNA web resources

Resource	Description	PMID
miRDB	MiRNA target predictions and functional annotations	18426918
miRNAMap	Experimental verified miRNAs and target genes	18029362
miRGen	Genomic locations and clusters of miRNAs; predicted and validated targets	17108354
miRGator	An integrated system for functional annotation of miRNAs	17942429
miRecords	Predicted and validated miRNA-target interactions in animals	18996891
mirBase	MiRNA sequence database	16381832
miR2Disease	Human miRNA diseases associations	18927107
microRNA.org	Predicted miRNA-target interactions and miRNA expression profiles	18158296
miRStart	MiRNA transcription start sites	21821656
MirMaid	Integrates miRNA data resources in a uniform web service interface	20074352
miRWalk	Predicted and validated miRNA targets	21605702
PhenomiR	MiRNA expression in diseases and biological processes	20089154
smirnaDB/mirZ	Expression profiles of miRNAs in animals	19468042
TargetRank	Ranking of conserved and non-conserved miRNA targets	17872505
MMIA	Web tool for examining biological functions of miRNA expression	19420067
ncRNAppi	Tool for identifying disease-related miRNA and siRNA targeting pathways	19808883
MiRror	A combinatorial analysis web tool for ensembles of miRNAs and their targets	20529892
PuTmiR	Putative miRNA transcription factors	20398296
TransmiR	Validated transcription factor-miRNA regulations	19786497
Magia	Integrates predictions from miRanda, PITA and TargetScan	20484379
miRò	MiRNA-phenotype associations in humans	20157481
TAM	Tool for annotations of miRNAs	20696049
FAME	Functional assignment of miRNAs via enrichment	20576699
SylArray	Automated detection of miRNA effects from expression data	20871108
MiRonTop	Integrates high-throughput transcriptomics data to identify the potential implication of miRNAs on a specific biological system	20959382
CircuitsDB	MiRNA-TF feed-forward loops in human and mouse	20731828
starBase	MiRNA-target interaction networks from Argonaute CLIP-Seq (HITS-CLIP) and degradome sequencing (Degradome-Seq, PARE) data	21037263
dbDEMC	Differentially expressed miRNAs in human cancers	21143814
miRTarBase	Experimentally verified miRNA target interactions	21071411
miRSel	Text-mining based identification of miRNA-target interactions	20233441

Resource	Description	PMID
CoGemiR	Conservation of miRNAs during evolution in different animal species	18837977
microPIR	Complementary miRNA seed sites in gene promotor regions	22439011
miRTrail	Regulatory interactions between genes and miRNAs in human diseases	22356618
miRSystem	Characterizing enriched functions and pathways of microRNA targets	22870325
OncomiRDB	Experimentally verified oncogenic and tumour-suppressive miRNAs	24651967
miRCancer	Literature-derived miRNA-cancer associations	23325619

Table B.1 Extensive list of miRNA web resources.
Beware that the entries are not structured according to the proposed classification. Furthermore, plant related resources are not listed.

C Details on miRNA target hub regulation

C.1 The *CDKN1A* regulatory network

miRNA	Target	Regulation	Evidence	Confidence score
hsa-miR-125a-5p	*CDKN1A*	repression	Wu *et al.* (2010)	0.734072303
hsa-miR-132	*CDKN1A*	repression	Wu *et al.* (2010)	0.734072303
hsa-miR-208a	*CDKN1A*	repression	Wu *et al.* (2010)	0.734072303
hsa-miR-28-5p	*CDKN1A*	repression	Wu *et al.* (2010)	0.800545314
hsa-miR-298	*CDKN1A*	repression	Wu *et al.* (2010)	0.734072303
hsa-miR-299-5p	*CDKN1A*	repression	Wu *et al.* (2010)	0.800545314
hsa-miR-345	*CDKN1A*	repression	Wu *et al.* (2010)	0.734072303
hsa-miR-363	*CDKN1A*	repression	Wu *et al.* (2010)	0.734072303
hsa-miR-423-3p	*CDKN1A*	repression	Wu *et al.* (2010)	0.847708666
hsa-miR-515-3p	*CDKN1A*	repression	Wu *et al.* (2010)	0.734072303
hsa-miR-572	*CDKN1A*	repression	Wu *et al.* (2010)	0.734072303
hsa-miR-639	*CDKN1A*	repression	Wu *et al.* (2010)	0.734072303
TF	**Target**	**Regulation**	**Evidence**	**Confidence score**
EGR1	miR-125a-5p	unknown	PMID: 20811575	0.859493492
PAX5	miR-125a-5p	unknown	predicted	0.235707303
NRSF	miR-125a-5p	unknown	predicted	0.235707303
ZID	miR-125a-5p	unknown	predicted	0.235707303
PPARG	miR-125a-5p	unknown	predicted	0.268706736
GATA3	miR-125a-5p	unknown	predicted	0.337665441
MZF1	miR-125a-5p	unknown	predicted	0.337665441
TP53	miR-125a-5p	unknown	predicted	0.337665441
NFIC	miR-125a-5p	unknown	predicted	0.337665441
EGR1	miR-132	activation	PMID: 20811575	0.92566572
CREB1	miR-132	activation	PMID: 17994010	0.859493492
AREB6	miR-132	unknown	predicted	0.268706736
E47	miR-132	unknown	predicted	0.235707303
ELK1	miR-132	unknown	predicted	0.268706736
EGR2	miR-132	unknown	predicted	0.292120206
FOS	miR-132	unknown	predicted	0.235707303
NFIC	miR-132	unknown	predicted	0.235707303
RELA	miR-132	unknown	predicted	0.268706736
MAX	miR-208a	unknown	predicted	0.268706736
EGR1	miR-208a	unknown	predicted	0.292120206
ARNT	miR-208a	unknown	predicted	0.268706736
RFX1	miR-208a	unknown	predicted	0.235707303
ATF6	miR-208a	unknown	predicted	0.235707303
YY1	miR-208a	unknown	predicted	0.268706736
JUNB	miR-208a	unknown	predicted	0.235707303
POU2F1	miR-208a	unknown	predicted	0.235707303
NFIC	miR-208a	unknown	predicted	0.268706736
MZF1	miR-28-5p	unknown	predicted	0.282657247
E47	miR-28-5p	unknown	predicted	0.268706736
FOXC1	miR-28-5p	unknown	predicted	0.310281105
FOXI1	miR-28-5p	unknown	predicted	0.235707303
SRY	miR-28-5p	unknown	predicted	0.268706736
JUN	miR-28-5p	unknown	predicted	0.292120206
POU2F1	miR-28-5p	unknown	predicted	0.235707303
SRF	miR-28-5p	unknown	predicted	0.268706736

CEBPA	miR-28-5p	unknown	predicted	0.292120206
RORA	miR-28-5p	unknown	predicted	0.235707303
JUNB	miR-298	unknown	predicted	0.310281105
CREB1	miR-299-5p	unknown	predicted	0.235707303
SRY	miR-299-5p	unknown	predicted	0.268706736
SRF	miR-299-5p	unknown	predicted	0.292120206
RUNX1	miR-299-5p	unknown	predicted	0.268706736
SP1	miR-345	unknown	predicted	0.337665441
RELA	miR-345	unknown	predicted	0.268706736
NFATC2	miR-345	unknown	predicted	0.235707303
HNF4A	miR-345	unknown	predicted	0.310281105
NFYA	miR-345	unknown	predicted	0.235707303
POU2F1	miR-345	unknown	predicted	0.235707303
NFIC	miR-345	unknown	predicted	0.310281105
E2F1	miR-363	activation	PMID: 19034270	0.615384615
FOXC1	miR-363	unknown	predicted	0.325119639
MZF1	miR-363	unknown	predicted	0.337665441
BACH1	miR-423-3p	unknown	predicted	0.268706736
STAT1	miR-423-3p	unknown	predicted	0.268706736
POU3F2	miR-423-3p	unknown	predicted	0.235707303
STAT5A	miR-423-3p	unknown	predicted	0.268706736
SRF	miR-423-3p	unknown	predicted	0.268706736
MEF2A	miR-423-3p	unknown	predicted	0.268706736
POU2F1	miR-423-3p	unknown	predicted	0.235707303
FOXF2	miR-572	unknown	predicted	0.235707303
NFYA	miR-639	unknown	predicted	0.235707303
YY1	miR-654-3p	unknown	predicted	0.268706736
BACH1	miR-654-3p	unknown	predicted	0.235707303
FOXL1	miR-654-3p	unknown	predicted	0.292120206
SRY	miR-654-3p	unknown	predicted	0.235707303
NFATC2	miR-654-3p	unknown	predicted	0.235707303
FOS	miR-654-3p	unknown	predicted	0.235707303
POU2F1	miR-654-3p	unknown	predicted	0.268706736
NFKB1	miR-654-3p	unknown	predicted	0.235707303
HNF1A	miR-654-3p	unknown	predicted	0.235707303
TFAP2A	miR-657	unknown	predicted	0.337665441
SP1	miR-657	unknown	predicted	0.337665441
MZF1	miR-657	unknown	predicted	0.337665441
MAX	miR-657	unknown	predicted	0.292120206
USF1	miR-657	unknown	predicted	0.268706736
EGR1	miR-657	unknown	predicted	0.337665441
RELA	miR-657	unknown	predicted	0.292120206
TFAP2A	miR-93	unknown	predicted	0.337665441
SOX9	miR-93	unknown	predicted	0.235707303
GATA3	miR-93	unknown	predicted	0.268706736
MZF1	miR-93	unknown	predicted	0.310281105
SP1	miR-93	unknown	predicted	0.337665441
POU2F1	miR-93	unknown	predicted	0.235707303
EGR1	miR-93	unknown	predicted	0.292120206
CEBPA	miR-93	unknown	predicted	0.235707303
NFYA	miR-93	unknown	predicted	0.235707303
RUNX1	miR-93	unknown	predicted	0.235707303
STAT1	miR-93	unknown	predicted	0.235707303

E2F1	miR-93	activation	PMID: 19034270	0.851091919
MYC	miR-93	activation	PMID: 20878079	0.841332593
SP1	*CDKN1A*	activation	PMID: 11064455, 9268365, 9405248, 9553123, 8557703	1
SP3	*CDKN1A*	activation	PMID: 11064455, 9268365 , 9405248, 9553123, 8557703	1
TFAP2A (AP2)	*CDKN1A*	activation	PMID: 8988173	0.596553373
STAT1	*CDKN1A*	activation	PMID: 8614832	0.724787961
STAT3	*CDKN1A*	activation	PMID: 9694869	0.659915801
STAT5A (STAT5)	*CDKN1A*	activation	PMID: 9111365	0.671565798
BRCA1	*CDKN1A*	activation	PMID: 9296497	0.596553373
TCF3 (E2A)	*CDKN1A*	activation	PMID: 9315646	0.724787961
VDR	*CDKN1A*	activation	PMID: 8566748	0.596553373
RARA (RAR)	*CDKN1A*	activation	PMID: 8940196	0.596553373
CEBPA (C/EBPα)	*CDKN1A*	activation	PMID: 9442037	0.596553373
CEBPB (C/EBPβ)	*CDKN1A*	activation	PMID: 9359698	0.596553373
TP53 (p53)	*CDKN1A*	activation	PMID: 8242752	0.671565798
TP63 (p63)	*CDKN1A*	activation	PMID: 15095006	0.671565798
TP73 (p73)	*CDKN1A*	activation	PMID: 15095006	0.671565798
RB1 (Rb)	*CDKN1A*	activation	PMID: 11314031	0.596553373
E2F1 (E2F)	*CDKN1A*	inhibition	PMID: 15048076	0.766070277
TBX2	*CDKN1A*	inhibition	PMID: 14996726	0.596553373
RUNX1	*CDKN1A*	inhibition	PMID: 10617663	0.766070277
RUNX2	*CDKN1A*	inhibition	PMID: 12391164	0.766070277
CUX1 (CDP/cut)	*CDKN1A*	inhibition	PMID: 15269344, 9707427	0.659915801
EHMT2 (G9a)	*CDKN1A*	inhibition	PMID: 15269344	0.596553373
JUN (v-jun)	*CDKN1A*	inhibition	PMID: 12717415	0.596553373
SRF	*CDKN1A*	unknown	predicted	0.432668931
NFKB1	*CDKN1A*	unknown	predicted	0.435893504
RELA	*CDKN1A*	unknown	predicted	0.435893504
STAT2	*CDKN1A*	unknown	predicted	0.437016231
STAT6	*CDKN1A*	unknown	predicted	0.437016231

PPI	PPI	Interaction type	# of publications	Confidence score
AKT1	CDKN1A	direct interaction	1	0.718340644
BCCIP	CDKN1A	direct interaction	2	0.809501577
C1orf123	CDKN1A	direct interaction	1	0.718340644
CASP3	CDKN1A	direct interaction	3	0.874181289
CCDC85B	CDKN1A	direct interaction	1	0.718340644
CCNA1	CDKN1A	direct interaction	3	0.874181289
CCNA2	CDKN1A	direct interaction	2	0.809501577
CCNB1	CDKN1A	direct interaction	1	0.718340644
CCNB2	CDKN1A	direct interaction	1	0.718340644
CCND1	CDKN1A	direct interaction	3	0.874181289
CCND2	CDKN1A	direct interaction	2	0.809501577
CCND3	CDKN1A	direct interaction	1	0.718340644
CCNE1	CDKN1A	direct interaction	1	0.718340644
CCNE2	CDKN1A	direct interaction	2	0.809501577
CDC20	CDKN1A	Complex	-	0.5625
CDC27	CDKN1A	unknown	-	0.5625
CDC45	CDKN1A	direct interaction	1	0.718340644
CDC5L	CDKN1A	direct interaction	1	0.718340644
CDC6	CDKN1A	direct interaction	1	0.718340644
CDC7	CDKN1A	direct interaction	1	0.718340644

CDK1	CDKN1A	direct interaction	1	0.718340644
CDK14	CDKN1A	direct interaction	1	0.718340644
CDK2	CDKN1A	direct interaction	1	0.718340644
CDK3	CDKN1A	direct interaction	1	0.718340644
CDK4	CDKN1A	direct interaction	1	0.718340644
CDK5	CDKN1A	unknown	-	0.5625
CDK6	CDKN1A	direct interaction	1	0.718340644
CEBPA	CDKN1A	direct interaction	1	0.718340644
CIZ1	CDKN1A	direct interaction	1	0.718340644
CSNK2A1	CDKN1A	direct interaction	1	0.718340644
CSNK2B	CDKN1A	direct interaction	1	0.718340644
DAPK3	CDKN1A	direct interaction	1	0.718340644
DHX9	CDKN1A	unknown	-	0.5625
ESR1	CDKN1A	direct interaction	1	0.718340644
GADD45A	CDKN1A	direct interaction	1	0.718340644
GADD45B	CDKN1A	direct interaction	1	0.718340644
GADD45G	CDKN1A	direct interaction	1	0.718340644
GMNN	CDKN1A	direct interaction	1	0.718340644
GNB2L1	CDKN1A	direct interaction	1	0.718340644
HDAC11	CDKN1A	direct interaction	1	0.718340644
ITGB1BP3	CDKN1A	direct interaction	1	0.718340644
MAP3K5	CDKN1A	direct interaction	1	0.718340644
MAPK8	CDKN1A	direct interaction	1	0.718340644
MCM10	CDKN1A	direct interaction	1	0.718340644
MDM2	CDKN1A	complex	-	0.5625
NR4A1	CDKN1A	unknown	-	0.5625
PARP1	CDKN1A	direct interaction	1	0.718340644
PCNA	CDKN1A	direct interaction	4	0.92435077
PIM1	CDKN1A	direct interaction	1	0.718340644
POLD2	CDKN1A	direct interaction	1	0.718340644
PRKCH	CDKN1A	unknown	-	0.5625
PSMA3	CDKN1A	direct interaction	1	0.718340644
RAB1A	CDKN1A	direct interaction	1	0.718340644
RNF144B	CDKN1A	direct interaction	1	0.718340644
RPA1	CDKN1A	unknown	-	0.5625
SET	CDKN1A	direct interaction	1	0.718340644
SKP2	CDKN1A	direct interaction	3	0.874181289
SLC25A11	CDKN1A	direct interaction	1	0.718340644
STAT3	CDKN1A	direct interaction	1	0.718340644
TEX11	CDKN1A	direct interaction	1	0.718340644
TK1	CDKN1A	direct interaction	1	0.718340644
TP53	CDKN1A	direct interaction	6	1
TSG101	CDKN1A	direct interaction	1	0.718340644
TTLL5	CDKN1A	direct interaction	1	0.718340644
XRCC6	CDKN1A	direct interaction	1	0.718340644

Table C.1 Molecular interactions in the *CDKN1A* regulatory network.

<table>
<tr><th>miRNA</th><th>Start/End</th><th>Position[1]</th><th>miRNA mature sequence</th><th>Predicted alignment</th></tr>
<tr><td>miR-208a</td><td>36653647/
36653669</td><td>69</td><td>AUAAGACGAGCAAAAAGCUUGU</td><td>miRNA 3'-uguucgaaAAACGAGCAGAAUa-5'
| | ||||: :||||||
mRNA 5'-agucucagUUUGUGUGUCUUAa-3</td></tr>
<tr><td>miR-298</td><td>36653618/
36653645</td><td>40</td><td>AGCAGAAGCAGGGAGGUUCUCCCA</td><td>miRNA 3'-acccUCUUGGAGGGA---CGAAGACGa-5'
| ||: || |:|| |||||||
mRNA 5'-ucaaAGGCCCGCUCUACAUCUUCUGCc-3'</td></tr>
<tr><td>miR-299-5p</td><td>36654982/
36655004</td><td>1404</td><td>UGGUUUACCGUCCCACAUACAU</td><td>miRNA 3'-uacauacaccCUGCCAUUUGGu-5'
||| |||||||
mRNA 5'-ucuccaccuaGACUGUAAACCu-3'</td></tr>
<tr><td>miR-299-5p</td><td>36654627/
36654649</td><td>1049</td><td>UGGUUUACCGUCCCACAUACAU</td><td>miRNA 3'-TA-TATGATGGGGGAGTAGATCT-5'
|||| ||||| |||:|||
mRNA 5'-UACAUAC-ACCCUGCCAUUUGGU-3'</td></tr>
<tr><td>miR-515-3p</td><td>36654029/
36654050</td><td>451</td><td>GAGUGCCUUCUUUUGGAGCGUU</td><td>miRNA 3'-TGCGAGGTTTTCTTCCGTGAG-5'
|| || ||| ||||||:
mRNA 5'-AAGT--AAACAGATGGCACTT-3'</td></tr>
<tr><td>miR-93</td><td>36654029/
36654051</td><td>451</td><td>CAAAGUGCUGUUCGUGCAGGUAG</td><td>miRNA 3' TGGACGT--GCTTGTCGTGAAA 5'
| |:| | :|::|||||||
Target 5' AGAAGTAAACAGATGGCACTTT 3'</td></tr>
<tr><td>miR-93</td><td>36654706/
36654732</td><td>1128</td><td>CAAAGUGCUGUUCGUGCAGGUAG</td><td>miRNA 3'-gauGGACGUG-CUUGU--CGUGAAAc-5'
| ||| | | | || ||||||||
mRNA 5'-aucCCUCCCCAGUUCAUUGCACUUUg-3</td></tr>
<tr><td>miR-125a-5p</td><td>36654604/
36654627</td><td>1026</td><td>UCCCUGAGACCCUUUAACCUGUGA</td><td>miRNA 3'-GTG-GGGTTATCTCTGTGTTAGGGG-5'
|||||| || ||||||:
mRNA 5'-AGUGUCCAAUUUCC-CAGAGUCCCU-3'</td></tr>
<tr><td>miR-132</td><td>36654928/
36654949</td><td>1350</td><td>UAACAGUCUACAGCCAUGGUCG</td><td>miRNA 3'-gcugguACCGACAUCUGACAAu-5'
|: |||| |:|||||||
mRNA 5'-acauacUGGC-CUGGACUGUUu-3'</td></tr>
<tr><td>miR-28-5p</td><td>36655034/
36655056</td><td>1456</td><td>AAGGAGCUCACAGUCUAUUGAG</td><td>miRNA 3'-gaguuaucugacacUCGAGGAa-5'
: :: :| |||||||
mRNA 5'-ucugugucuuucacAGCUCCUc-3'</td></tr>
<tr><td>miR-28-5p</td><td>36654906/
36654928</td><td>1328</td><td>AAGGAGCUCACAGUCUAUUGAG</td><td>miRNA 3'-gaGUUAUCU-GACACUCGAGGAa-5'
||| | | || |||||||
mRNA 5'-ggCAA ACACCUUCCAGCUCCUg-3'</td></tr>
<tr><td>miR-345</td><td>36654735/
36654757</td><td>1157</td><td>GCUGACUCCUAGUCCAGGGCUC</td><td>miRNA 3'-cucggGACCUGAUCCUCAGUCg-5'
: | | || |||||||||
mRNA 5'-agcagCGGAACAAGGAGUCAGa-3'</td></tr>
<tr><td>miR-654-3p</td><td>36654738/
36654761</td><td>1160</td><td>UAUGUCUGCUGACCAUCACCUU</td><td>miRNA 3'-uuccacuaccagUC-GUCUGUAu-5'
|: | | || |||||||
mRNA 5'-agcggaacaaggAGUCAGACAUu-3'</td></tr>
</table>

<table>
<tr><th>miRNA</th><th>Start/End</th><th>Position[1]</th><th>miRNA mature sequence</th><th>Predicted alignment</th></tr>
<tr><td>miR-363</td><td>36654358/
36654378</td><td>776</td><td>AAUUGCACGGUAUCCAUCUGUA</td><td>miRNA 3'-au--GUCuACCuaUGGcACGUUAA-5'
||| ||| ||| |||||||
mRNA 5'-tctcCAGcTGG--GCTcTGCAATT-3'</td></tr>
<tr><td>miR-423-3p</td><td>36654053/
36654074</td><td>470</td><td>AGCUCGGUCUGAGGCCCCUCAGU</td><td>miRNA 3' ugAC-UCCCCGGAGUCUGGCUCGa 5'
|| ||||||||| ||||||:
Target 5' ttTGAAGGGGCCTC--ACCGAGTg 3'</td></tr>
<tr><td>miR-423-3p</td><td>36654138/
36654162</td><td>560</td><td>AGCUCGGUCUGAGGCCCCUCAGU</td><td>miRNA 3' ugACUC--CCCGGAGUCUGGCUCGa 5'
|||| |||| :|:|:||||
Target 5' agTGAGCACAGCCTAGGGCTGAGCt 3'</td></tr>
<tr><td>miR-423-3p</td><td>36654277/
36654304</td><td>699</td><td>AGCUCGGUCUGAGGCCCCUCAGU</td><td>miRNA 3' ugACUCCCCG----GAGUC-UGGCUCGa 5'
||:|| || ||||| :::||||
Target 5' acTGGGGAGCCCGTCTCAGTGTTGAGCc 3'</td></tr>
<tr><td>miR-572</td><td>36654730/
36654749</td><td>1152</td><td>GUCCGCUCGGCGGUGGCCCA</td><td>miRNA 3' acccgguggcggcUCGCCUg 5'
||||||
Target 5' cactttgattagcAGCGGAa 3'</td></tr>
<tr><td>miR-639</td><td>36654522/
36654549</td><td>944</td><td>AUCGCUGCGGUUGCGAGCGCUGU</td><td>miRNA 3' ugUCGCGAGCGUUGG---CGUCGCUa 5'
| | | |||:|: |||||||
Target 5' cuACCUCAGGCAGCUCAAGCAGCGAc 3'</td></tr>
<tr><td>miR-657</td><td>36653580/
36653602</td><td>2</td><td>GGCAGGUUCUCACCCUCUCUAGG</td><td>miRNA 3' ggaucucucccaCUCUUGGACGg 5'
|: |:|||||
Target 5' aatccgcccacaGGAAGCCTGCa 3'</td></tr>
<tr><td>miR-657</td><td>36654250/
36654272</td><td>672</td><td>GGCAGGUUCUCACCCUCUCUAGG</td><td>miRNA 3' ggaucucucccacucUUGGACGg 5'
:||||||
Target 5' cctcatggcccctctGACCTGCa 3'</td></tr>
</table>

[1] Location in the 3'UTR of the target mRNA

Table C.2 MiRNA binding sites in the *CDKN1A* 3′ UTR.

Survey toward a confidence scoring system for molecular interaction data

We carried out a survey to determine appropriate values for weighting the scores that are used to define the confidence in each interaction in the *CDKN1A* regulatory network. The survey was conducted among a panel of experts in the fields of network biology and miRNA regulation, experimentalists and modellers, from six different laboratories listed below. Precisely, we elaborated the survey in which we asked individually to each member of the expert panel to assign values to the weights and scores ($k_{<p,t,bs,r,cons,pred,m>}$ and S_m) for the different interaction types considered in our network. The answers were collected, averaged and rounded for simplicity (Table C.3). The resulting values were used to establish our confidence scoring system described in Chapter 3. Laboratories participating in this survey were: the Department of Systems

Biology and Bioinformatics (University of Rostock; three participants), the Department of Dermatology (University of Leipzig; one participant), the Institute of Experimental Gene Therapy and Cancer Research (University of Rostock; one participant), the Indian Institute of Toxicology Research (Lucknow, India; one participant), the Department of Molecular Pulmonology (Philipps University Marburg; one participant), and the Institute for Biostatistics and Informatics in Medicine and Ageing Research (University of Rostock, one participant).

Scores	weights	SP1	SP2	SP3	SP4	SP5	SP6	SP7	SP8	Ø
PPIs (S_{PPI})	k_P	0.7	0.8	0.8	0.8	0.7	0.7	0.7	0.7	**0.7**
	k_T	0.9	0.9	1	0.8	0.9	0.9	0.9	0.9	**0.9**
TF-*CDKN1A* (S_{TF})	k_P	0.7	0.8	0.8	0.8	0.7	0.7	0.7	0.7	**0.7**
	k_T	0.9	0.8	0.9	0.8	0.9	0.9	0.9	0.9	**0.9**
	k_BS	0.9	0.9	0.9	0.9	0.8	0.9	0.9	0.9	**0.9**
	k_Cons	0.3	0.35	0.4	0.3	0.5	0.3	0.3	0.3	**0.3**
	k_BS	0.9	0.5	0.7	0.6	0.8	0.9	0.9	0.9	**0.8**
miR-*CDKN1A* (S_{miR})	k_P	0.7	0.8	0.7	0.7	0.7	0.7	0.7	0.7	**0.7**
	k_M	1	0.7	1	1	1	1	1	1	**1**
	k_BS	0.4	0.5	0.8	0.6	0.8	0.4	0.4	0.4	**0.5**
TF-miR (S_{TFmiR})	k_P	0.9	0.8	0.7	0.8	0.7	0.9	0.9	0.9	**0.8**
	k_Pred	0.6	0.4	0.4	0.4	0.5	0.6	0.6	0.6	**0.5**
	k_n	1	1	1	1	1	1	1	1	**1**
	k_BS	0.7	0.7	0.7	0.7	0.7	0.7	0.7	0.7	**0.7**
Methods (S_M)	reporter assay	0.8	0.8	0.8	0.8	0.8	0.9	0.8	0.8	**0.8**
	western blot	0.6	0.75	0.7	0.7	0.7	0.9	0.6	0.6	**0.7**
	qRT-PCR	0.4	0.6	0.5	0.4	0.5	0.3	0.4	0.4	**0.4**
	microarray	0.25	0.4	0.4	0.2	0.5	0.3	0.25	0.25	**0.3**

Table C.3 Results of the survey for establishing a confidence scoring scheme.
The table shows the results of the survey that was conducted among domain experts for the development of a confidence scoring scheme for molecular interaction data (SP = survey participant).

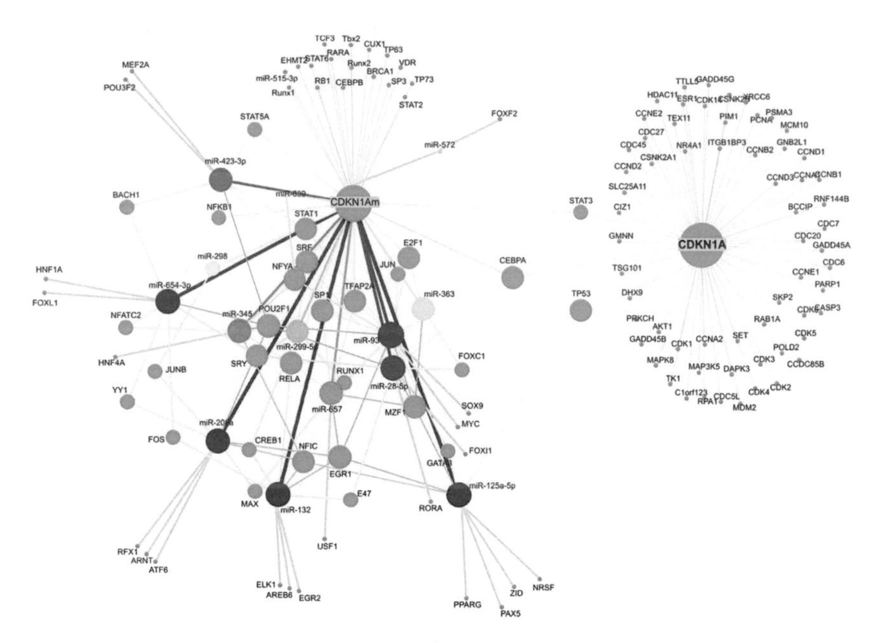

Figure C.1 Topology of the *CDKN1A* regulatory network.
The size of the nodes is determined by their degree (the number of linked edges). Therefore, *CDKN1A* mRNA as well as the CDKN1A protein are the largest nodes in this network representation. The colours of the nodes reflect their betweenness centrality value, which represents the amount of control that a node exerts over the interactions of other nodes in the network (low values: green; high values: red). Interestingly, some miRNAs have a high betweenness centrality as compared to others, which may suggest a more central (important) role of these entities in the regulation of *CDKN1A* expression. Furthermore, we found that that these miRNAs are regulated by a number of TFs and that they are involved in several ($n > 2$) FFLs (not shown).The figure was generated with Cytoscape.

C.2 Modelling *CDKN1A* target hub regulation

Model construction

We constructed a kinetic model of miRNA target regulation and used experimental data from Wu *et al.* (2010) to characterise the parameter values. The kinetic model accounts for the evolution in time of the mRNA ($CDKN1A_m$) and protein ($CDKN1A_p$) expression levels of the miRNA target hub *CDKN1A*, the *CDKN1A*-targeting miRNAs (miR_i; $i = 1 \dots 15$) verified by Wu *et al.* (2010), and the complexes formed by target mRNA and miRNA, $[CDKN1A_m \,|\, miR_i]$. In total, the model includes 32 time-dependent variables and 64 parameters:

$$\frac{d}{dt}CDKN1A_m = k_{syn\_CDKN1A_m} \cdot f_{act}(TF_{CDKN1A_m}) - CDKN1A_m \cdot \left(k_{deg\_CDKN1A_m} + \sum_i k_{ass\_miR_i} \cdot miR_i\right)$$

$$\frac{d}{dt}miR_i = k_{syn\_miR_i} \cdot f_{act}(TF_{miR_i}) - miR_i \cdot k_{deg\_miR_i} - k_{ass\_miR_i} \cdot CDKN1A_m \cdot miR_i$$

$$\frac{d}{dt}[CDKN1A_m|miR_i] = k_{ass\_miR_i} \cdot CDKN1A_m \cdot miR_i - k_{deg\_comp_i} \cdot [CDKN1A_m|miR_i]$$

$$\frac{d}{dt}CDKN1A_p = k_{syn\_CDKN1A_p} \cdot CDKN1A_m - k_{deg\_CDKN1A_p} \cdot CDKN1A_p$$

In the model the following processes are considered. For the *CDKN1A* mRNA ($CDKN1A_m$): (i) basal synthesis (characterised by the parameter $k_{syn\_CDKN1Am}$), which is mediated by its transcription factors ($f_{act}(TF_{CDKN1Am})$); (ii) basal degradation ($k_{deg\_CDKN1Am}$); and (iii) association with an miRNA (k_{ass_miRi}). For each *CDKN1A*-targeting miRNA (miR_i): (i) basal synthesis (k_{syn_miRi}) mediated by its transcriptional factors ($f_{act}(TF_{miRi})$); (ii) basal degradation (k_{deg_miRi}); and (iii) association with the *CDKN1A* mRNA target (k_{ass_miRi}). For each complex ($[CDKN1A_m | miR_i]$): (i) association of miR_i and *CDKN1A* mRNA to form a complex (k_{ass_miRi}); and (ii) degradation (k_{deg_compi}). For CDKN1A protein ($CDKN1A_p$): (i) mRNA-mediated synthesis of protein ($k_{syn\_CDKN1Ap}$); and (ii) degradation ($k_{deg\_CDKN1Ap}$). According to our model, the total measurable amounts for *CDKN1A* mRNA and each *CDKN1A*-targeting miRNA are defined by the equations:

$$CDKN1A_{m\_tot} = CDKN1A_m + \sum_i [CDKN1A_m|miR_i]$$

$$miR_{i\_tot} = miR_i + \sum_i [CDKN1A_m|miR_i]$$

To characterise the potential cooperative effect associated with the proximity of miRNA binding sites (Figure 3.4), we defined a group of new variables ($[CDKN1A_p|miR_i|miR_j]$), which account for ternary complexes composed of *CDKN1A* mRNA and two potential cooperating miRNAs (miR_i and miR_j). For these new variables, two processes are considered: (i) the association of *CDKN1A* mRNA with miR_i and miR_j into the complex ($k_{dass\_miRi,j}$); and (ii) the degradation of the complex ($k_{deg\_compi,j}$). The expanded model reads as follows:

$$\frac{d}{dt}CDKN1A_m = k_{syn\_CDKN1A_m} \cdot f_{act}(TF_{CDKN1A_m}) - CDKN1A_m \cdot \left(k_{deg\_CDKN1A_m} + \sum_i k_{ass_{miR_i}} \cdot miR_i + \sum_{i,j} k_{dass_{miRi},j} \cdot miR_i \cdot miR_j \right)$$

$$\frac{d}{dt}miR_i = k_{syn\_miR_i} \cdot f_{act}(TF_{miR_i}) - miR_i \cdot (k_{deg\_miRi} + CDKN1A_m \cdot (k_{ass_{miR_i}} + \sum_j k_{dass\_miRi,j} \cdot miR_j))$$

$$\frac{d}{dt}[CDKN1A_m|miR_i] = k_{ass\_miR_i} \cdot CDKN1A_m \cdot miR_i - k_{deg\_comp_i} \cdot [CDKN1A_m|miR_i]$$

$$\frac{d}{dt}[CDKN1A_m|miR_i|miR_j] = k_{ass\_miR_{i,j}} \cdot CDKN1A_m \cdot miR_i \cdot miR_j - k_{deg\_comp_{i,j}} \cdot [CDKN1A_m|miR_i|miR_j]$$

$$\frac{d}{dt}CDKN1A_p = k_{syn\_CDKN1A_p} \cdot CDKN1A_m - k_{deg\_CDKN1A_p} \cdot CDKN1A_p$$

After including these new variables, the equations for the total amount of *CDKN1A* mRNA and each free miR_i also have to be reformulated:

$$CDKN1A_{m\_tot} = CDKN1A_m + \sum_i [CDKN1A_m|miR_i] + \sum_j [CDKN1A_m|miR_i|miR_j]$$

$$miR_{i\_tot} = miR_i + \sum_i [CDKN1A_m|miR_i] + \sum_j [CDKN1A_m|miR_i|miR_j]$$

Calibration of the CDKN1A model

For calibrating the model, the parameter values and initial conditions of the state variables were set and characterised using the following strategy. The model was normalised around the basal, non-miRNA repressed, levels of *CDKN1A* mRNA and protein (th_m and th_p were assumed to be 1 when no miRNA repression occurs). Additionally, experimental data describing their half-lives were used to assign values to their basal degradation (k_{deg_thm} and k_{deg_thp}) and synthesis (k_{syn_thm}, k_{syn_thp}) rates (Maki and Howley, 1997; Wang *et al.*, 2000). Recent reports indicated that miRNA half-life can vary under different cellular contexts (Bhattacharyya *et al.*, 2006; Hwang *et al.*, 2007; Krol *et al.*, 2010). However, in our analysis we assumed a constant degradation rate ($k_{deg\_miRi} = 0.0289$) due to lack of experimental data that can be used to characterise cell context dependent degradation. This assumption is supported by the evidence that miRNAs are rather stable molecules with half-lives of more than 24 hours (Kai and Pasquinelli, 2010).

Furthermore, because *CDKN1A* expression can be regulated by two different miRNA silencing mechanisms we formulated distinctive assumptions for each mechanism

(Filipowicz *et al.*, 2008). In case of translation repression, the degradation rate of *CDKN1A* mRNA is not affected by miRNA and we, therefore, assumed $k_{deg\_compi} = k_{deg\_thm}$. In case of mRNA deadenylation followed by degradation, as this mechanism accelerates the loss of target mRNA, we assumed $k_{deg\_compi} > k_{deg\_thm}$. We extracted quantitative data from Wu *et al.* (2010), which describes *CDKN1A* mRNA and protein levels upon single miRNA overexpression, and used them to estimate parameters in our model via data fitting (Table C.4). The *CDKN1A* mRNA and protein levels were measured 48 hours after transfection of 100 nM of single *CDKN1A* targeting miRNAs to HEK 293 cells. Because of the high amount of transfected miRNA, we assumed that the endogenous production of these miRNAs is negligible ($k_{syn\_miRi} = 0$). Subsequently, the rest of the parameters were estimated by using an iterative method combining global (particle swarm pattern search algorithm; Vaz and Vicent 2007) and local (downhill simplex method in multidimensions; Press *et al.*, 1993) optimisation algorithms. The method minimises the distance between model simulations and experimental data using the following cost function:

$$F_{error} = \frac{1}{n_{exp} \cdot n_{prot} \cdot n_{tpoints}} \sum_{k=1}^{n_{exp}} \sum_{i=1}^{n_{prot}} \sum_{h=1}^{n_{tpoints}} \left[\left(mRNA_{i,k}(\tau_h) - mRNA_{i,k}^{exp}(\tau_h) \right)^2 + \left(Prot_{i,k}(\tau_h) - Prot_{i,k}^{exp}(\tau_h) \right)^2 \right]$$

where n_{exp} is the number of experiments; n_{prot} is the number of miRNA regulated proteins; $n_{tpoints}$ is the number of time points considered in each experiment; $mRNA_{i,k}(\tau)$ represents the simulated value of mRNA i at time τ in the experiment k; $mRNA_{i,k}^{exp}(\tau)$ stands for the experimentally measured value for the mRNA i at time τ in the experiment k; $Prot_{i,k}(\tau)$ is the simulated value of protein i level at time τ and $Prot_{i,k}^{exp}(\tau)$ the experimentally measured value for protein i at time τ in the experiment k. Wu and colleagues (2010) measured the CDKN1A protein and mRNA expression levels 48 hrs after individual overexpression of each miRNA. Due to the experimental design used by Wu *et al.* (2010), the parameter estimation problem can be divided into 15 simpler problems, whose cost function reduces to the following:

$$F_{error,miRi} = \left[th_m(\tau_{48}) - th_m^{exp}(\tau_{48}) \right]^2 + \left[th_p(\tau_{48}) - th_p^{exp}(\tau_{48}) \right]^2$$

In each one of these problems, data from transfection of miR_i from Wu *et al.* (2010) is used to estimate the values of the parameters k_{ass_miRi} and k_{deg_compi}. Ultimately, after the model calibration procedure we obtained the values for the estimated parameters that minimise the residuals (F_{error}, miR_i) (Table C.4). As shown in Figure C.2, the characterised model correctly predicts most experimental data. However, in case of data from transfection experiments with miRNAs namely miR-298, miR-208a, miR-132 and miR-28-5p the agreement between model and data is not perfect. In such cases mRNA expression after transfection is higher than in the control case. Given its structure, our model cannot capture this behaviour, which may be caused by experimental noise or unknown regulatory mechanisms. Moreover, a stronger repression of the target gene is observed when two miRNAs bind to the target mRNA at proximate sites (Doench and Sharp, 2004). Thus, we assumed a stronger association rate for the complex $[th_m|miR_i|miR_j]$ which is equal to the aggregation of their individual association rates $(k_{ass\_miRi,j} = k_{ass\_miRi} + k_{ass\_miRj})$. Similarly, the degradation rate of the complex $[th_m|miR_i|miR_j]$ was assumed to be the aggregation of degradation rates of single miRNA binding complexes $(k_{deg\_compi,j} = k_{deg\_compi} + k_{deg\_compj})$.

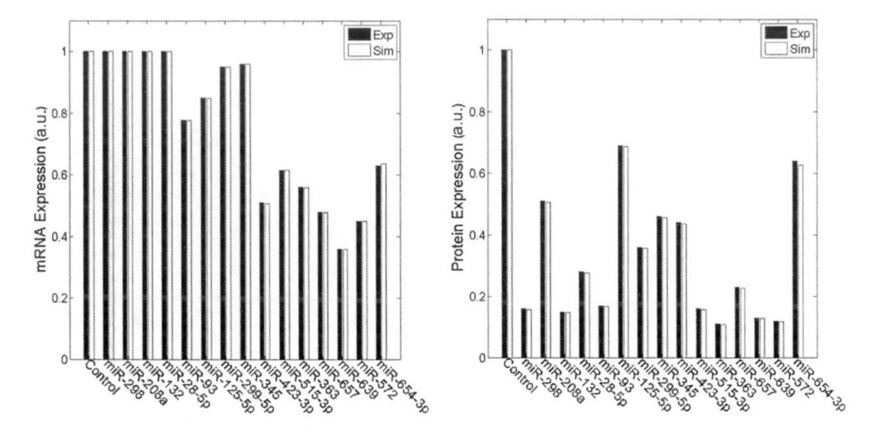

Figure C.2 Fitting the model to the experimental data in Wu *et al.* (2010).
The model parameters are numerically calculated using computational algorithms that minimise the distance between model simulations (Sim) and experimental data (Exp). We show the comparison between both quantities for the chosen set of parameter values in Table C.4. **Left**: The measured expression levels of *CDKN1A* mRNA after overexpression of the miRNAs as indicated are compared to the model predictions. The values are normalised to the scenario (Control) in which miRNAs are normally expressed. **Right**: The measured expression levels of CDKN1A protein after overexpression of the miRNAs as indicated are compared to the model predictions. The values are normalised to the scenario (Control) in which miRNAs are normally expressed (a.u.: arbitrary unit).

Parameter	Description	Value	Comment
k_{syn_thm}	Basal synthesis of *CDKN1A* mRNA	0.1155 hr^{-1}	assumed
k_{deg_thm}	Basal degradation of *CDKN1A* mRNA	0.1155 hr^{-1}	Wang *et al.*, 2000
$k_{ass\_miRi}, (i = 1 \dots 15)$	Association of the complex [CDKN1Am/miRNA$_i$]	[0.001 10] (n.u.· hr)$^{-1}$	estimated
$k_{deg\_compi}, (i = 1 \dots 15)$	Basal degradation of complex [miRNA/mRNA$_i$]	0.1155 hr^{-1} or [0.001 10] hr^{-1}	assumed or estimated
$k_{syn\_miRi}, (i = 1 \dots 15)$	Basal synthesis of miRNA$_i$	0.0289 hr^{-1}	assumed
$k_{deg\_miRi}, (i = 1 \dots 15)$	Basal degradation of miRNA$_i$	0.0289 hr^{-1}	Kai and Pasquinelli 2010
k_{syn_thp}	Basal synthesis of CDKN1A protein	1.3863 hr^{-1}	assumed
k_{deg_thp}	Basal degradation of CDKN1A protein	1.3863 hr^{-1}	Maki and Howley 1997

Initial conditions and TF functions

Variable	Description	Value (a.u.)
$th_p(0)$	Protein level of CDKN1A	1
$th_m(0)$	mRNA level of *CDKN1A*	1
$[th_m \mid miR_{i=1 \dots 15}](0)$	The complex miRNA$_i$/mRNA	0
$miR_{i=1 \dots 15}(0)$	CDKN1A-targeting miRNA$_i$	100
$f_{act}(TF_{thm})$	Activation function of *CDKN1A* TFs	1
$f_{act}(TF_{miRi\ (i=1 \dots 15)})$	Activation function of miRNA TFs	1

Estimated parameter values for each *CDKN1A*-targeting miRNA

miRNA$_i$	k_{deg_compi} (i=1...15)	k_{ass_miRi} (i=1...15)	Residuals ($F_{error,\ miRi}$)	Experimental data (CDKN1A protein)	Experimental data (*CDKN1A* mRNA)
miR-298 (miR_1)	0.1155	0.0254	3.4e-004	0.16	1.074
miR-208a (miR_2)	0.1155	0.0041	2.0e-003	0.51	1.192
miR-132 (miR_3)	0.1155	0.0275	2.4e-003	0.15	1.210
miR-28-5p (miR_4)	0.1155	0.0119	5.9e-003	0.28	1.350
miR-125-5p (miR_5)	0.1155	0.0018	1.8e-003	0.69	0.850
miR-299-5p (miR_6)	0.1155	0.0080	1.8e-004	0.36	0.950
miR-345 (miR_7)	0.1155	0.0051	1.1e-004	0.46	0.960
miR-93 (miR_8)	0.1564	0.0235	4.1e-014	0.17	0.778
miR-423-3p	0.9118	0.0055	2.8e-009	0.44	0.510

(miR_9)					
miR-515-3p (miR_{10})	0.2098	0.0253	1.2e-013	0.16	0.616
miR-363(miR_{11})	0.2261	0.0399	2.2e-014	0.11	0.560
mR-657 (miR_{12})	0.3465	0.0158	2.1e-014	0.23	0.480
miR-639 (miR_{13})	0.4305	0.0327	1.8e-017	0.13	0.360
miR-572 (miR_{14})	0.3039	0.0360	9.4e-023	0.12	0.450
miR-654-3p (miR_{15})	9.7485	0.0024	3.0e-014	0.64	0.630

Table C.4 Model definition and parameters.
The parameters are classified into three categories: (1) parameters whose values are obtained from the publications as mentioned in the table; (2) parameters whose values are fixed in a way the steady-state levels of CDKN1A protein and mRNA are normalised to 1 in absence of miRNA repression, in accordance with the data in Wu *et al.* (2010); (3) parameters whose values are estimated by data fitting within a biologically feasible interval. The initial concentration of *CDKN1A* mRNA and protein are set to 1, which is equal to the value of their steady-state levels when no miRNA/mRNA interactions are present. The initial concentration of *CDKN1A* targeting miRNAs is set to 100, which is consistent with the experimental setting used by Wu *et al.* (2010). For simplicity, the TF functions are assumed to be 1. Based on the experimental observations, the *CDKN1A*-targeting miRNAs verified in Wu *et al.* (2010) are divided into two groups: the translation repression group (blue shaded cells) and the mRNA deadenylation group (orange shaded cells). Considering usual values of noise in biological data, a miRNA is classified into the mRNA deadenylation group if after miRNA overexpression the *CDKN1A* mRNA level is repressed by 20% or more (mRNA level $\leq$ 0.8, the control level is 1), otherwise it is classified into the translation repression group. For the translation repression group, only the parameter k_{ass_miRi} is estimated, while $k_{deg\_compi} = k_{deg\_thm} = 0.1155$ (complex mRNA/miR assumed to have the same half-life as a single mRNA). For the mRNA deadenylation group, both k_{ass_miRi} and k_{deg_compi} are estimated.

C.3 Simulations of context-dependent *CDKN1A* expression

Three hypothetical target regulation mechanisms

For validating the hypothesis of cooperative target regulation by miRNAs with binding sites in close proximity and the effect on target response to external stimuli, we carried out overexpression experiments of miR-572 and miR-93, individually and in combination, in cells under genotoxic stress. The binding sites of these two miRNAs were predicted to be located in close proximity in the 3' UTR of *CDKN1A* mRNA (see Figure 3.4; the distance is 24 nt) and could facilitate cooperative target repression. The experiments were performed to measure the CDKN1A protein expression under genotoxic stress in four scenarios (Figure C.6A): (1) both miRNAs are normally expressed; (2) miR-572 is overexpressed, miR-93 is normally expressed; (3) miR-572 is normally expressed, miR-93 is overexpressed; (4) both miRNAs are moderately overexpressed.

We validated the predictive ability of our model (Appendix C.3) using these experimental data. In the model we characterised the function for the transcriptional activation of *CDKN1A* ($f_{act}(TF_{CDKN1Am})$) by using the experimental data that describes the response of TP53 after genotoxic stress induction (Figure C.6B). TP53 is typically up-regulated upon DNA damage and is known as transcriptional activator of *CDKN1A* (el-Deiry *et al.*, 1994; Sax and El-Deiry, 2003). To mimic the four experimental settings, we set the corresponding parameter values as follows:

(1) $miR_8(0) = miR_{14}(0) = 1, k_{syn\_miR8} = k_{deg\_miR8}, k_{syn\_miR14} = k_{deg\_miR14}$

(2) $miR_8(0) = miR_{14}(0) = 1, k_{syn\_miR8} = k_{deg\_miR8}, k_{syn\_miR14} = 100 \cdot k_{deg\_miR14}$

(3) $miR_8(0) = miR_{14}(0) = 1, k_{syn\_miR8} = 100 \cdot k_{deg\_miR8}, k_{syn\_miR14} = k_{deg\_miR14}$

(4) $miR_8(0) = miR_{14}(0) = 1, k_{syn\_miR8} = 50 \cdot k_{deg\_miR8}, k_{syn\_miR14} = 50 \cdot k_{deg\_miR14}$

In agreement with the experimental design, we did not modulate the initial concentration of miRNA but increased their synthesis rates. As shown in Figure 3.6, the model predictions correspond well to the experimental observations of *CDKN1A* response after genotoxic stress induction in the four different miRNA expression scenarios. More specifically, when both miRNAs are normally expressed, CDKN1A protein expression is most strongly up-regulated due to the DNA damage stimulus. Overexpression of miR-572 or miR-93 is able to reduce the up-regulation of CDKN1A stimulated by genotoxic stress. Finally, simultaneous overexpression of both miRNAs shows the most effective suppression of the stress-induced up-regulation of CDKN1A. This effect is even enhanced by the cooperative action of the two miRNAs. Based on these results, we conclude that the workflow for the detection of cooperative binding sites is valid for *CDKN1A*-targeting miRNAs, and therefore, it can also be applied to detect potential cooperative sites for other miRNA target hubs.

We further used miR-93 (miR_8) and miR-572 (miR_{14}) as an example to illustrate the detailed dynamics of the three hypothetical miRNA regulatory mechanisms by which CDKN1A protein expression (CDKN1A*p*) might be modulated. For each target regulation mechanism, the corresponding parameter values are set as:

i) Independent target regulation

$$k_{ass\_miR8} = 0.0235,\ k_{ass\_miR14} = 0.036,\ k_{dass\_miR8,14} = 0$$

ii) Interdependent target regulation

$$k_{ass\_miR8} = 0,\ k_{ass\_miR14} = 0,\ k_{dass\_miR8,14} = k_{ass\_miR8} + k_{ass\_miR14} = 0.0595$$

iii) Synergistic target regulation

$$k_{ass\_miR8} = 0.0235,\ k_{ass\_miR14} = 0.036,\ k_{dass\_miR8,14} = k_{ass\_miR8} + k_{ass\_miR14} = 0.0595$$

Subsequently, we simulated the model by changing the initial concentrations of both miRNAs ($miR_8(0)$ and $miR_{14}(0)$) at an interval ranging from 10^{-1} to 10^2 and computed the expression levels of CDKN1A protein at 48 hrs. In this way, our simulation setup is comparable with the experimental results generated by Wu *et al.* (2010), in which the CDKN1A protein and mRNA levels are measured at 48 hrs after miRNA transfection. We plotted the CDKN1A expression levels for different combinations of initial miRNA concentrations in Figure 3.5.

In order to investigate how the different strengths of miRNA cooperativity can affect CDKN1A protein expression (CDKN1A*p*), we simulated the synergistic target regulation of CDKN1A by miR-93 and miR-572 for four different initial concentration scenarios ($[miR_8(0), miR_{14}(0)]$): (1) both miRNAs are normally expressed ($[1,1]$); (2) miR-93 is normally expressed, miR-572 is over expressed ($[1,10]$); (3) miR-93 is over expressed, miR-572 is normally expressed ($[10,1]$); (4) both miRNAs are over expressed. For each scenario, we defined an interval accounting for the strength of miRNA cooperativity ($K = k_{dass\_miR8,14} \cdot [10^{-5}\ 10^{5}]$). Finally, we simulated the model, assigning values of K in steps of $10^{0.25}$, and for every round of simulation we computed the expression levels of CDNK1A protein at 48 hrs as shown in Figure 3.8.

CDKN1A regulation in different cell biological processes

We used our model to predict the CDKN1A steady-state levels for different cell biological processes and to analyse the role of miRNA cooperativity. We first extracted the list of GO terms associated with the miRNA transcription factors (TFs). We then assumed that miRNA expression is activated when at least one of the miRNA TFs is associated with the biological process under investigation. For example, TP53 is a TF for miR-125a-5p and it is associated with cell proliferation, so we assumed miR-125a-5p to be active ($miR_5(0) = 1$) when cells are proliferating. In contrast, we assumed miR-125a-5p to be non-active ($miR_5(0) = 0$) in DNA repair because none of its TFs is associated with that biological process. By following this strategy, the miRNA expression profiles for each of the nine cell biological processes were obtained (Table C.5).

To compute the CDKN1A steady-states levels for each biological process, we set the following parameter values and initial concentrations for the model: (1) for each biological process, the initial conditions of the miRNAs ($miR_{i=1...15}(0)$) were derived from the Table C.5; (2) the initial concentrations for CDKN1A protein and mRNA were set to one ($CDKN1A_m(0) = CDKN1A_p(0) = 1$) and the synthesis rate of each miRNA was set to zero ($k_{syn\_miRi(i=1...15)} = 0$); and (3) when considering miRNA cooperativity, the corresponding parameters values ($k_{ass\_miRi,j}$ and $k_{deg\_compi,j}$) were assigned according to Table C.6. Otherwise, these parameters were set to zero ($k_{dass\_miRi,j} = 0, k_{deg\_compi,j} = 0$). For each biological process, we configured the model as discussed and computed the steady-state levels of CDKN1A (Figure 3.7).

miRNA	Proliferation	Apoptosis	Cell Cycle	Immune Response	Inflamm. Response	Senescence	DNA Damage	Migration
miR-298 (miR_1)	1	0	1	0	0	0	0	0
miR-208a (miR_2)	1	1	1	1	0	0	1	0
miR-132 (miR_3)	1	1	1	1	1	0	0	0
miR-28-5p (miR_4)	1	1	1	1	0	0	0	0
miR-125a-5p (miR_5)	1	1	1	1	0	1	1	0
miR-299-5p (miR_6)	0	1	0	1	0	1	0	1
miR-345 (miR_7)	1	1	1	1	1	0	1	0
miR-93 (miR_8)	1	1	1	0	0	0	0	0
miR-423-3p (miR_9)	1	1	1	1	1	1	0	1
miR-515-3p (miR_{10})	0	0	0	0	0	0	0	0
miR-363(miR_{11})	1	1	1	0	0	0	0	0
mR-657 (miR_{12})	1	1	0	1	1	0	0	0
miR-639 (miR_{13})	0	0	0	0	0	0	0	0
miR-572 (miR_{14})	0	0	0	0	0	0	0	0
miR-654-3p (miR_{15})	1	1	0	1	1	0	1	0

Table C.5 Expression profiles of *CDKN1A*-targeting miRNAs for different biological processes.

Variable [CDKN1Am|miRi|miRj]	Parameter $k_{dass\_miRi,j}$	$k_{deg\_Compi,j}$	Initial concentration
$[CDKN1A_m\|miR_3\|miR_4]$	$k_{ass\_miR3} + k_{ass\_miR4}$	$k_{deg\_comp3} + k_{deg\_comp4}$	0
$[CDKN1A_m\|miR_5\|miR_6]$	$k_{ass\_miR5} + k_{ass\_miR6}$	$k_{deg\_comp5} + k_{deg\_comp6}$	0
$[CDKN1A_m\|miR_4\|miR_{11}]$	$k_{ass\_miR4} + k_{ass\_miR11}$	$k_{deg\_comp4} + k_{deg\_comp11}$	0
$[CDKN1A_m\|miR_1\|miR_2]$	$k_{ass\_miR1} + k_{ass\_miR2}$	$k_{deg\_comp1} + k_{deg\_comp2}$	0
$[CDKN1A_m\|miR_1\|miR_{11}]$	$k_{ass\_miR1} + k_{ass\_miR11}$	$k_{deg\_comp1} + k_{deg\_comp11}$	0
$[CDKN1A_m\|miR_7\|miR_8]$	$k_{ass\_miR7} + k_{ass\_miR8}$	$k_{deg\_comp7} + k_{deg\_comp8}$	0
$[CDKN1A_m\|miR_8\|miR_{15}]$	$k_{ass\_miR8} + k_{ass\_miR15}$	$k_{deg\_comp8} + k_{deg\_comp15}$	0
$[CDKN1A_m\|miR_8\|miR_9]$	$k_{ass\_miR8} + k_{ass\_miR9}$	$k_{deg\_comp8} + k_{deg\_comp9}$	0
$[CDKN1A_m\|miR_8\|miR_{14}]$	$k_{ass\_miR8} + k_{ass\_miR14}$	$k_{deg\_comp8} + k_{deg\_comp14}$	0
$[CDKN1A_m\|miR_9\|miR_{10}]$	$k_{ass\_miR9} + k_{ass\_miR10}$	$k_{deg\_comp9} + k_{deg\_comp10}$	0
$[CDKN1A_m\|miR_9\|miR_{12}]$	$k_{ass\_miR9} + k_{ass\_miR12}$	$k_{deg\_comp9} + k_{deg\_comp12}$	0

Table C.6 Variables and parameters accounting for the miRNA cooperativity.

Tissue-specific CDKN1A expression profiles

First, we extracted the miRNA expression levels in different tissues from the database miRNAMap (release 2.0; Hsu *et al.*, 2006) and normalized them according to the most abundant miRNA observed in each tissue. For example, in adipose the most abundant miRNA is miR-125a-5p. The expression of miR-125a-5p is normalized to 1 ($miR_5(0) = 1$), and accordingly other miRNA expression levels are rescaled. If a miRNA does not appear in a tissue expression profile its expression level is consider as 0 (Table C.7). Second, for each tissue we computed the steady-state levels of *CDKN1A* by changing the initial concentrations of the miRNAs according to their normalized tissue-specific expression levels, while other parameters were configured as described in Table C.4 and Table C.6. Third, we obtained the experimental data of *CDKN1A* expression levels in different tissues from the ArrayExpress database (version as of January 2012; Brazma, 2003). The tissue-specific *CDKN1A* expression levels were discretized as low or high according to the experimental evidence from the database. For example, in adipose and kidney the *CDKN1A* expression levels were observed to be over- and underexpressed respectively, so we can say that the *CDKN1A* expression level in kidney is lower than in adipose. Out of 12 tissues, only in kidney and liver *CDKN1A* was underexpressed. Thus, in kidney and liver the *CDKN1A* expression level was categorized as low while for the other tissues it was categorized as high (Table C.7). Finally, we compared the model predictions with the experimental data for the *CDKN1A* expression levels. As shown in Figure C.3, the model is able to correctly predict the relative *CDKN1A* expression levels in 9 out of 12 tissues.

miRNA	kidney	liver	adipose	bladder	intestine	cervix	heart	lung	ovary	testis	trachea	brain
miR-298	0	0	0	0	0	0	0	0	0	0	0	0
miR-208a	0	0	0	0	0	0	0	0	0	0	0	0
miR-132	0.051	0.033	0.071	0.053	0.069	0.018	0.04	0.008	0.03	0.033	0.07	0.134
miR-28-5p	0.012	0.02	0.021	0.05	0.025	0.008	0.015	0.008	0.012	0.006	0.03	0.002
miR-125a-5p	1	0.5	1	1	0.462	1	1	1	1	1	0.9	1
miR-299-5p	0.001	0.001	0.001	0.001	0.002	0.002	0.001	$1 \cdot 10^{-4}$	0.002	0.004	0.001	0.002
miR-345	0.005	0.005	0.004	0.002	0.004	0.001	0.001	0.001	0.003	0.003	0.007	0.001
miR-93	0.314	0.417	0.357	0.433	0.577	0.112	0.15	0.302	0.275	0.167	0.5	0.104
miR-423-3p	0	0	0	0	0	0	0	0	0	0	0	0
miR-515-3p	0	0	0	0	0	0	0	0	0	0	0	0
miR-363	0.005	0.001	0.001	0.001	0.002	0.001	$3 \cdot 10^{-4}$	0.001	0.001	0.001	0.003	0.001
mR-657	$2 \cdot 10^{-4}$	0.001	0.001	$2 \cdot 10^{-4}$	0.001	$1 \cdot 10^{-4}$	$3 \cdot 10^{-4}$	$1 \cdot 10^{-4}$	$2 \cdot 10^{-4}$	0.001	0.001	$2 \cdot 10^{-4}$
miR-639	0	0	0	0	0	0	0	0	0	0	0	0
miR-572	0.001	0.005	0.021	0.006	0.012	0.002	0.001	0.002	0.001	0.003	0.011	0.001
miR-654-3p	0	0	0	0	0	0	0	0	0	0	0	0
CDKN1A	**low**	**low**	**high**	**high**	**high**	**high**	**high**	**high**	**high**	**high**	**high**	**high**

Table C.7 The normalised tissue-specific miRNA and CDKN1A expression profiles. MiRNA expression profiles were retrieved from the miRNAMap database (release 2.0; Hsu *et al.*, 2006). *CDKN1A* expression levels (low: underexpression; high: overexpression) were extracted from the ArrayExpress database (version as of January 2012; Brazma, 2003).

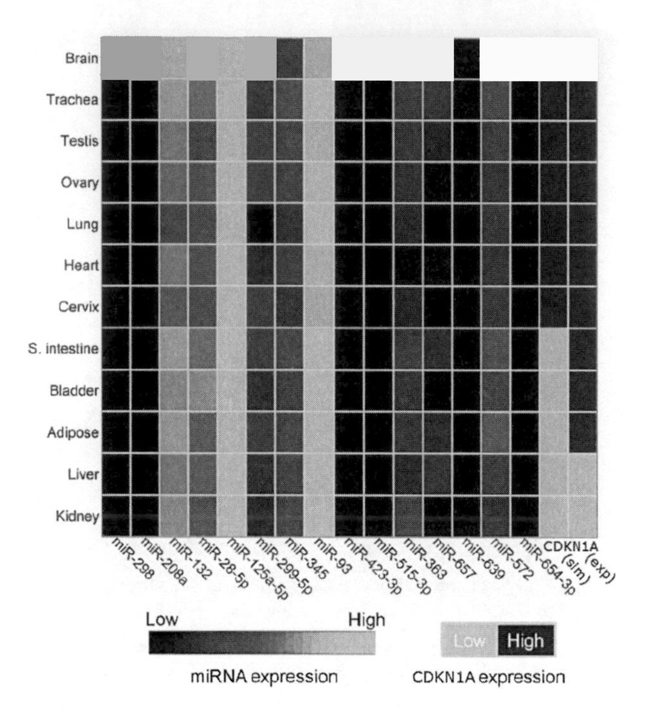

Figure C.3 Predicted tissue-specific *CDKN1A* expression profiles.

The tissue-specific miRNA expression profiles obtained from the database miRNAMap were normalised and used in the model to calculate the tissue-specific *CDKN1A* expression profiles (sim). The model correctly predicts the *CDKN1A* expression for 9 out of 12 tissues as compared to the experimental data obtained from the database ArrayExpress (exp).

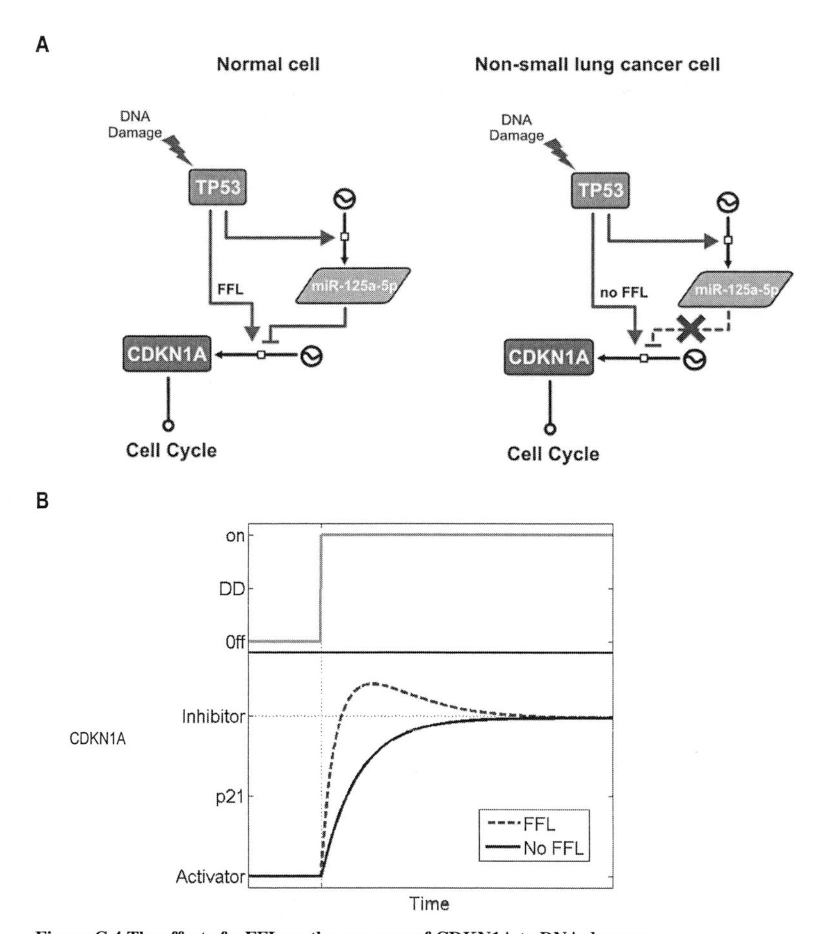

Figure C.4 The effect of a FFL on the response of CDKN1A to DNA damage.
(**A**) Upon DNA damage, TP53 gets stabilised and promotes the expression of proteins and miRNAs related to cell cycle arrest and apoptosis, including CDKN1A (DNA damage→TP53→genes; el-Deiry *et al.*, 1994). The level of DNA damage and subsequent TP53 activation determine the role of CDKN1A as an activator or an inhibitor during the cell cycle process by modulating its expression (Jung *et al.*, 2010). The response of CDKN1A to DNA damage are determined by two conditions: in normal cells, the feed forward loop formed by TP53, CDKN1A and miR-125a-5p is activated due to the normal expression of miR-125a-5p (left); in non-small lung cancer cells, for example, the feed-forward loop formed by TP53, CDKN1A and miR-125a-5p is deactivated due to the down-regulation of miR-125a-5p (right; Jiang *et al.*, 2010a). (**B**) In the simulation, we assume a step-like activation of DNA damage (DD). The simulations show that the response of CDKN1A was quicker and stronger when the feed-forward loop is activated. This result suggests that the suppression of the feed-forward loop mediated by miR-125a-5p (dashed line) can favour cancer progression by delaying the initiation of CDKN1A-triggered cell cycle arrest in response to DNA damage.

The response of CDKN1A to upstream stimulus signals

For analysing the effect of miRNA regulation on *CDKN1A*, we defined a new term (S) for *CDKN1A* stimulation by a transient stimulus signal, which is a function of the amplitude (μ) and duration (τ) of the signal. In our model, this term was added to the differential equation accounting for *CDKN1A* mRNA ($CDKN1A_m$). We further assumed non-basal synthesis for *CDKN1A* mRNA ($k_{syn\_CDKN1Am} = 0$), and set initial conditions to $CDKN1A_m(0) = 0$ and $CDKN1A_p(0) = 0$. In this manner, the CDKN1A response is exclusively activated by the transient stimulus signal. After introducing the new term, the differential equation for $CDKN1A_m$ becomes:

$$\frac{d}{dt}CDKN1A_m = S(\mu,\tau) - CDKN1A_m \cdot (k_{deg\_CDKN1Am} + \sum_i k_{ass\_miRi} \cdot miR_i + \sum_{i,j} k_{dass\_miRi,j} \cdot miR_i \cdot miR_j)$$

Subsequently, in order to keep consistent with experimental conditions in Wu *et al.* (2010) we assumed non-basal synthesis for each miRNA ($k_{syn\_miRi(i=1...15)} = 0$). Furthermore, we defined five miRNA abundance scenarios by setting the initial concentrations for the miRNAs and corresponding parameters accordingly:

(1) ***off***: the concentrations of all *CDKN1A*-targeting miRNAs are zero ($miR_{i=1...15}(0) = 0$, $k_{dass\_miRi,j} = 0$, $k_{deg\_compi,j} = 0$);

(2) ***on***: all miRNAs are at a normally expressed ($miR_{i=1...15}(0) = 1$, $k_{dass\_miRi,j} = 0$, $k_{deg\_compi,j} = 0$);

(3) ***on+C***: normally expressed miRNAs, plus cooperativity for miRNAs with binding sites in close proximity ($miR_{i=1...15}(0) = 1$, see Table C.6 for $k_{dass\_miRi,j}$ and $k_{deg\_compi,j}$);

(4) ***on(x10)***: miRNAs are 10-fold upregulated compared to normal ($miR_{i=1...15}(0) = 10$, $k_{dass\_miRi,j} = 0$, $k_{deg\_compi,j} = 0$);

(5) ***on(x10)+C***: 10-fold upregulated miRNAs, plus cooperativity ($miR_{i=1...15}(0) = 10$, see Table C.6 for $k_{dass\_miRi,j}$ and $k_{deg\_compi,j}$).

Firstly, we plotted the peak of the CDKN1A response to transient stimulus signals for the five miRNA abundance scenarios (Figure C.5B). The simulations were performed for different settings of the stimulus signal amplitude and duration ($\mu \in [10^{-2}\ 10^{1}]$ a.u. and $\tau = [1,10,24]$ hr). The maximum value of CDKN1A protein concentration for each simulation was computed, which was considered as the peak of the CDKN1A response. Next, we compared the dynamics of the CDKN1A response when the *CDKN1A* mRNA synthesis is activated by a signal with low ($\mu = 0.1$ a.u.) or high ($\mu = 10$ a.u.) amplitude, and for long-lasting ($\tau = 10\ hr$) signal duration. To do so, we computed the time-series response of CDKN1A for the five miRNA abundance scenarios (Figure C.5C).

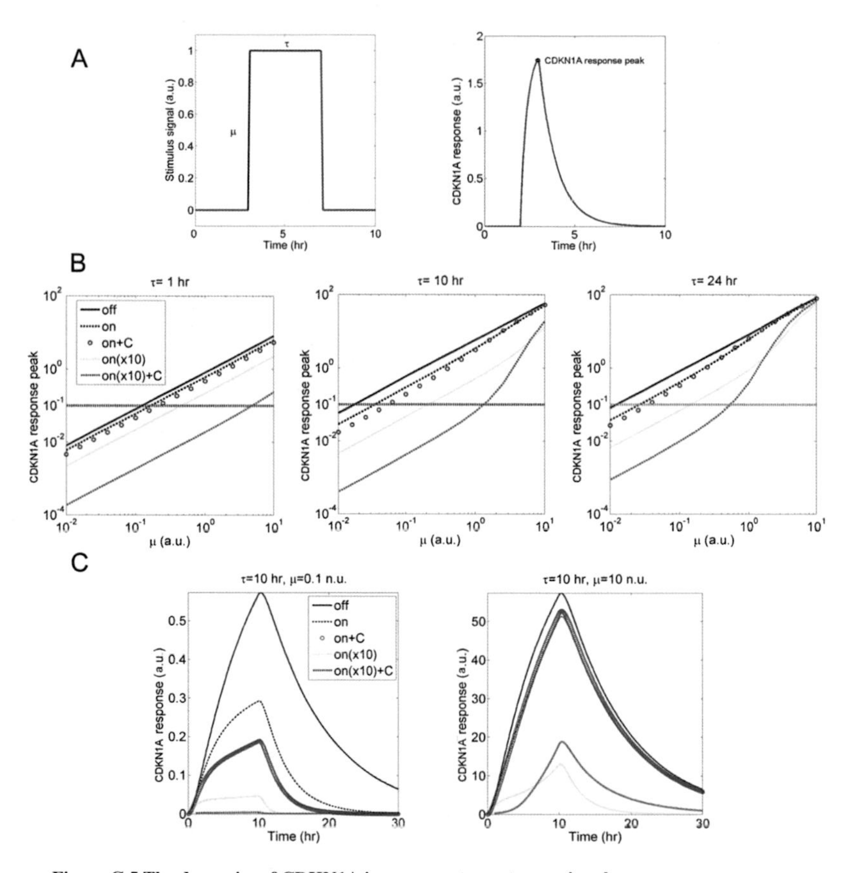

Figure C.5 The dynamics of CDKN1A in response to upstream signals.
A: The plot on the left shows a characteristic transient input signal which is characterized by the two variables μ (amplitude) and τ (duration). The plot on the right illustrates a hypothetical CDKN1A response over time to a certain input. **B:** The three plots show predicted peak values of CDKN1A in response to different input signals and different miRNA abundance scenarios. In all three plots the signal amplitude μ is in the range $[10^{-2}\ 10^{1}]$ while the CDKN1A response peaks are colored in red (no miRNA expression), black (normal miRNA expression), blue circles (miRNAs+cooperativity), light green (10-fold up-regulated miRNAs), and purple (10-fold up-regulated miRNAs+cooperativity). The difference between the three plots is the signal duration τ (left: 1 hr; middle: 10 hr; right: 24 hr). The grey dashed line marks the CDKN1A level which corresponds to 10% of the basal CDKN1A expression. C: These two plots illustrate the CDKN1A response over time 0-30 hr for the case of long-lasting stimuli ($\tau = 10\ hr$) with a low signal amplitude ($\mu = 0.1$; left) and a high signal amplitude ($\mu = 1$; right). For the signal amplitude we used arbitrary units (a.u.).

C.4 Experimental methods & results

Small RNA transfection and doxorubicin treatment

Sk-Mel-147 cells were seeded in a six well plate. Mature miRNA mimics (miR-572 #C-300891-01-0005, miR-93 #C-300512-07-0005 or Mimic Negative Control #1-CN-001000-01-05; Thermo Scientific, Dharmacon RNAi Technologies) were transfected individually at a concentration of 100 nM or in combination at 50 nM each using lipofectamine RNAimax reagent (Invitrogen, Karlsruhe, Schwerte, Germany). 48 hours after transfection the cells were pulse treated with 250 nM doxorubicin-HCl (Sigma-Aldrich, Munich, Germany) for 1 hour after which normal growth medium was replenished. The protein lysates were collected at 0, 2, 4, 6, 8 and 24 hours post-doxorubicin treatment.

Immunoblotting

At indicated time points the cells were harvested in radioimmunoprecipitation buffer [50 mmol/l Tris-HCL (pH 7.4), 150 mmol/L NaCl, 1% NP-40, 0.5% deoxycholic acid, 0.1% SDS], supplemented with protease inhibitor mix (Roche Molecular Biochemicals, Mannheim, Germany). 20 μg of total protein extract was denatured in electrophoresis sample buffer for 5 min at 95°C, and subjected to SDS-polyacrylamide gel electrophoresis (PAGE). Gels were electroblotted onto nitrocellulose membranes (Highbond ECLTM, Amersham, Braunschweig, Germany). Immunodetection was performed using an anti-CDKN1A mouse monoclonal Ab (sc-6246) and were re-probed with an anti-β-actin mouse monoclonal Ab (sc-5274) to verify equal loading of proteins (both the antibodies purchased from Santa Cruz Biotechnologies, Santa Cruz, CA, USA). The IRDye 800CW donkey antimouseIgG (926-32212) secondary antibody was used for immunodetection (LI-COR Biosciences, Bad Homburg, Germany). All the blots were processed in parallel and developed for 2 minutes. Immunodetection of TP53 was performed using a mouse monoclonal antibody against TP53 (#554293, BD Biosciences, Heidelberg, Germany) and the same secondary antibody as used for CDKN1A. The relative CDKN1A protein expressions are represented as fold changes compared to the CDKN1A expression in the nontargeting control at 0 hour time point. For detection and quantification of the band intensities we used LiCor Odyssey Western Blot detection system with integrated Image Studio software.

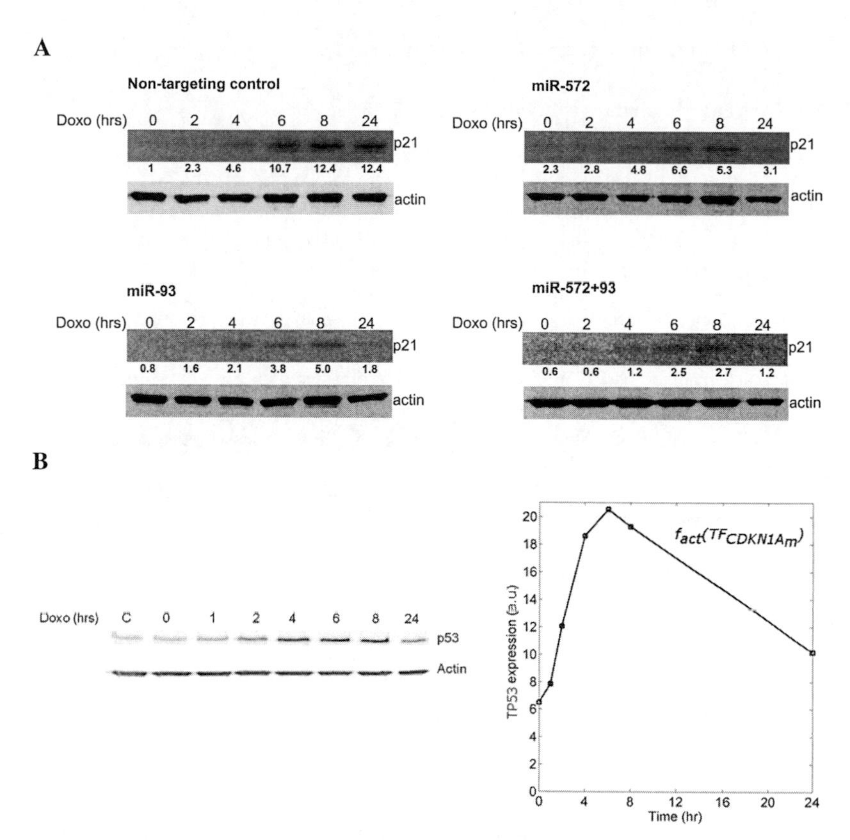

Figure C.6 Regulation of CDKN1A expression by miR-572 and miR-93.
(**A**) Western blots of CDKN1A protein levels after genotoxic stress. The CDKN1A expression (here referred to as p21) was measured at the indicated time points in four scenarios: (1) endogenous miRNA expression (non-targeting control); (2) only miR-572 overexpressed (miR-572); (3) only miR-93 overexpressed (miR-93); and (4) both miRNAs partially overexpressed (miR-572+93). The results suggest that miR-572 and miR-93 are able to repress CDKN1A not only individually but also in a cooperative manner when both of them are expressed simultaneously. (**B**) Western blot of TP53 (p53) protein levels at the indicated time points after genotoxic stress (left). The TP53 expression was used to characterise the transcriptional activation function of *CDKN1A* mRNA ($f_{act}(TF_{CDKN1Am})$; right).

D Details on cooperative target regulation

D.1 Molecular dynamics simulations

After structure editing, the charmm27 force field was assigned to the RNA complexes, which is a superset of the charmm22 force field with the additional coverage for nucleic acids (Foloppe and MacKerell, Jr., 2000). The nucleic acids parameters of the charmm27 force field have been successfully used in many molecular dynamics simulation studies in the past (Hart *et al.*, 2005; Banáš *et al.*, 2009; Paliy *et al.*, 2009).

The geometry of the initial 3D structure of RNA complexes was optimized using the Smart Minimizer protocol available in Accelrys® Discovery Studio 3.5 to remove steric overlap that produces bad contacts. The Smart Minimizer was run for a maximum of 5,000 steps with the Minimization RMS Gradient tolerance of 0.1 kcal/(mol x Å) to exit from the minimization routine in case the average gradient is less than or equal to the tolerance.

Molecular dynamics simulation setup

MDS were performed using the Simulation protocol available with Accelrys® Discovery Studio 3.5. The optimized structure of the complexes was heated gradually from 50 K to 300 K by scaling the velocity of each atom in a total of 10,000 steps with the iteration time step of 1 fs in order to prepare the system for the production run. After the heating phase, equilibration was performed to stabilize the system around the target temperature of 300 K by periodically reassigning velocities to each atom. The initial velocity of all the atoms was taken from the Maxwellian distribution at temperature 300 K by employing LeapFrog Verlet algorithms with the time step of 1 fs for 10,000 steps. RNA complexes were verified for their stability after the equilibration phase. For stable complexes, MD production simulations were performed for initially 100 ps duration. The canonical thermodynamic ensemble, i.e. constant-temperature and constant-volume ensemble (NVT) was selected for the production run. The motions of the molecules were recorded in the form of an output trajectory at each 1,000 steps. Hydrogen bond monitors were applied to the trajectory frames to investigate the stability of mRNA and miRNAs interactions during the course of the simulation. The production phase for stable complexes after 100 ps of initial run was further extended for another 400 ps to check the stability of the RNA complexes.

All the simulation steps were carried out in the Generalized Born with a simple SWitching (GBSW) implicit solvent model for the better approximation of the solvent effect on the complex. GBSW has been made computationally less expensive by replacing its computationally expensive molecular surface approximation with a van der Waals-based surface with a smooth dielectric boundary (Im *et al.*, 2003). The implicit solvent dielectric constant was set to 80. Lower and higher cut-off distance for non-bonded interactions were set to 10 and 12 Å. The non-bonded lists were maintained for the atom pairs within the distance of 14 Å beyond which the non-bonded interactions were ignored. Input atomic radii were taken from (Nina *et al.*, 1997) as recommended for the charmm27 force field. SHAKE constraint was applied to fix all bonds involving hydrogen bonds.

The role of Argonaute proteins in RNA triplexes

In our case study of cooperative target regulation in human we built 3D models of RNA triplexes composed of two cooperating miRNAs and a mutual target mRNA. However, a more realistic representation that would be closer to the native miRNA-induced target regulation would include the RISC complex which is crucial for miRNA-induced target repression. The most critical RISC component is the Argonaute protein which accommodates the miRNA molecule and coordinates target repression (Meister, 2013). However, with each detail that is added to the model or the simulations the computational cost increases significantly. Moreover, the crystal structure of Argonaute in complex with a miRNA was so far determined for only one case (hAgo2 + miR-20a). We used this complex structure which was determined by Elkayam and colleagues (Elkayam *et al.*, 2012; PDB ID: 4F3T) to generate, as proof of principle, a 3D model of cooperative target regulation with a participation of Argonaute proteins. Therefore, we searched our TriplexRNA database for targets cooperatively regulated by pairs of miR-20a molecules. We found two putative targets, namely *NEURL1B* and *ZBTB24*. We selected the former for the purpose of 3D model construction and MDS. More specifically, we constructed three models: (a) miR-20a/hAgo2 bound to the first binding site in the *NEURL1B* mRNA; (b) miR-20a/hAgo2 bound to the second binding site; and (c) one miR-20a/hAgo2 complex bound to each of the two neighbouring binding sites in the *NEURL1B* mRNA. See Figure D.1 for an illustration of all three 3D models.

The retrieved models indicate that the complex with two units miR-20a/hAgo2 hybridized with *NEURL1B* is more stable (potential energy: $-61{,}822.58\, kcal/mol$) than both complexes with a single unit of miR-20a/hAgo2 (potential energy: $-34{,}129.67\, kcal/mol$ and $-23{,}869.30\, kcal/mol$). Furthermore, the model of the RNA triplex (including two AGO proteins) exhibits an interaction between the two AGO proteins providing further stability to the complex (Figure D.1c).

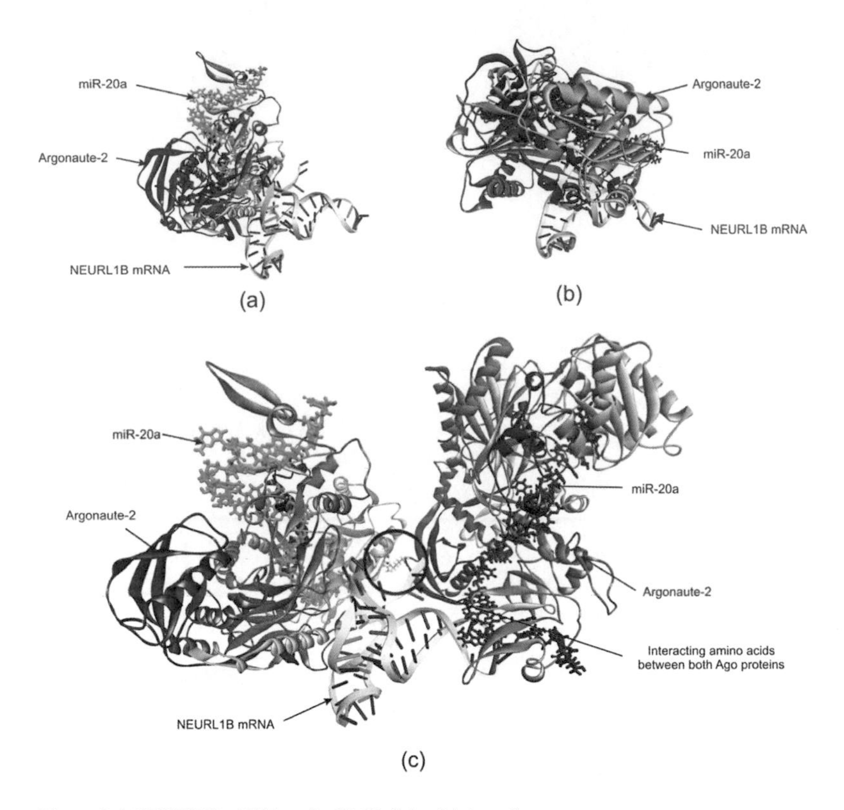

Figure D.1 ***NEURL1B*** **mRNA and miR-20a/hAgo2 interactions.**
The mRNA of the *NEURL1B* gene contains two binding sites for the miRNA miR-20a that reside in close proximity. Therefore a cooperative target repression is possible. The figure illustrates three putative complex structures that involve *NEURL1B* mRNA, miR-20a and the RISC protein hAgo2 which is essential in RNAi processes. **(a)** Shows a 3D model of an miR-20a/hAgo2 complex hybridized to the first miR-20a binding site in *NEURL1B*. **(b)** Shows a 3D model of an miR-20a/hAgo2 complex hybridized to the second miR-20a binding site in *NEURL1B*. **(c)** Shows a 3D model of two miR-20a/hAgo2 complexes hybridized to the first and second miR-20a binding site in *NEURL1B* respectively. In this last case two amino acid residues (THR555-ARG110) interact between the two AGO proteins (see black circle). The two miR-20a molecules are represented as green (1st binding site) and red (2nd binding site) stick models. Both AGO proteins are shown as solid ribbon and amino acid residues are colored in a continuous gradient from blue at the N-terminus through white to red at the C-terminus. The *NEURL1B* mRNA is illustrated with a yellow backbone and ladder shaped base pairs.

D.2 Parameter values of cooperative target regulation models

	$k_{ass}^{duplex_1}$	$k_{dis}^{duplex_1}$	$k_{ass}^{duplex_2}$	$k_{ass}^{triplex}$	$k_{dis}^{duplex_2}$	$k_{dis}^{triplex}$
ZNF121	0.0022	0.2016	0.0022	7.2e-06	0.2016	0.1665
HTRA2	0.0070	1.0000	0.0070	2.9e-05	1.0000	0.3904
RPS6KA5	0.0004	0.2647	0.0007	4.3e-06	0.2407	0.1902
ABT1	0.0830	0.1088	2.2e-06	0.9170	0.1265	0.0949
MUC1	0.0000	0.1271	0.0164	0.9836	0.1334	0.0897
EDA2R	0.0005	0.1093	0.0003	0.9991	0.1087	0.0782
C19orf69	0.9505	0.5011	0.0003	0.0001	0.5011	0.3478
KAT2B	0.8225	0.3047	0.0012	0.0001	0.2999	0.2516
NPHP1	0.2112	1.0000	0.0008	2.7e-05	0.9146	0.6524
GOLM1	0.0006	0.2342	0.0164	0.9830	0.2715	0.1560
CCDC3	0.0005	0.2500	1.3e-05	0.9995	0.2410	0.1878
PLXNB1	0.0000	0.2314	3.4e-07	0.9999	0.4699	0.1814

Table D.1 Parameter values of the kinetic model for selected cases.
These parameter values were used for simulating gene repression by cooperative miRNAs as shown in Figure 4.6.

D.3 Results of pathway enrichment analyses

Term	Count	%	PValue	Genes	Fold enrichm.	BH
Prostate cancer	13	5.16	1.16E-07	*E2F3, RELA, TP53, FOXO1, RAF1, PTEN, CCNE2, CCNE1, CDKN1A, HSP90B1, EP300, CDKN1B, BCL2*	7.35	1.29E-05
Pathways in cancer	22	8.73	9.92E-07	*E2F3, MSH2, TGFBR1, RELA, MITF, TP53, FOXO1, RAF1, PTEN, CCNE2, CCNE1, HSP90B1, PTK2, CDKN1A, EP300, HIF1A, CDKN1B, BCL2, VEGFA, RHOA, AXIN2, MYC*	3.38	5.51E-05
Cell cycle	12	4.76	2.82E-05	*CCNE2, CCNE1, CDKN1A, E2F3, CDKN1B, EP300, E2F5, TP53, YWHAE, MYC, ATM, WEE1*	4.83	1.04E-03
Small cell lung cancer	10	3.97	3.32E-05	*CCNE2, CCNE1, E2F3, PTK2, CDKN1B, RELA, BCL2, TP53, MYC, PTEN*	5.99	9.22E-04
TGF-beta signalling pathway	10	3.97	4.41E-05	*PPP2R1A, EP300, E2F5, SMAD7, TGFBR1, SMAD5, RHOA, ID4, MYC, ACVR1*	5.79	9.79E-04
Bladder cancer	7	2.78	1.47E-04	*RPS6KA5, CDKN1A, E2F3, VEGFA, TP53, RAF1, MYC*	8.39	2.72E-03
Chronic myeloid leukemia	8	3.17	6.12E-04	*CDKN1A, E2F3, CDKN1B, RELA, TGFBR1, TP53, RAF1, MYC*	5.37	9.66E-03
p53 signalling pathway	7	2.78	2.03E-03	*CCNE2, CCNE1, CDKN1A, PPM1D, TP53, PTEN, ATM*	5.18	2.79E-02

Term	Count	%	PValue	Genes	Fold enrichm.	BH
Neurotrophin signalling pathway	9	3.57	2.84E-03	*RPS6KA5, IRAK1, RELA, BCL2, RHOA, TP53, CALM3, RAF1, YWHAE*	3.65	3.45E-02
Colorectal cancer	7	2.78	5.87E-03	*MSH2, BCL2, TGFBR1, TP53, RAF1, AXIN2, MYC*	4.20	6.33E-02
Glioma	6	2.38	7.59E-03	*CDKN1A, E2F3, TP53, CALM3, RAF1, PTEN*	4.79	7.40E-02
Melanoma	6	2.38	1.24E-02	*CDKN1A, E2F3, MITF, TP53, RAF1, PTEN*	4.25	1.09E-01
Pancreatic cancer	6	2.38	1.32E-02	*E2F3, RELA, TGFBR1, VEGFA, TP53, RAF1*	4.20	1.07E-01
Endometrial cancer	5	1.98	1.85E-02	*TP53, RAF1, AXIN2, MYC, PTEN*	4.84	1.38E-01
ErbB signalling pathway	6	2.38	2.76E-02	*CDKN1A, PTK2, CDKN1B, ERBB3, RAF1, MYC*	3.47	1.87E-01
Apoptosis	6	2.38	2.76E-02	*IRAK1, RELA, BCL2, IL1RAP, TP53, ATM*	3.47	1.87E-01
Wnt signalling pathway	7	2.78	7.59E-02	*CSNK1A1, PPP2R1A, EP300, RHOA, TP53, AXIN2, MYC*	2.33	4.21E-01
Focal adhesion	8	3.17	9.95E-02	*PTK2, BCL2, VEGFA, RHOA, RAF1, COL5A3, PTEN, PXN*	2.00	4.96E-01

Table D.2 Pathway enrichment analysis results for triplexes with experimental support.

Term	Count	%	PValue	Genes	Fold Enrichm.	BH
Regulation of actin cytoskeleton	35	2.37	4.24E-05	*FGFR2, ITGAL, FGFR1, FGFR4, PDGFB, DIAPH1, FGF14, MRAS, PIP5K1B, ARPC4, RDX, ITGB3, VCL, PAK7, ITGB8, RHOA, RRAS, PIK3CA, PIK3R5, MSN, PAK1, CSK, FGF3, LIMK1, PIK3CD, ITGA3, MYL12A, NCKAP1, ITGA5, CFL2, ARAF, PDGFRA, CYFIP2, CRK, PIP4K2C*	2.09	7.35E-03
Pathways in cancer	41	2.78	2.48E-03	*E2F1, FGFR2, HSP90AB1, FGFR1, WNT5B, PDGFB, FGF14, PGF, PML, TCF7L2, WNT1, BCL2, PAX8, SLC2A1, CASP8, RHOA, PIK3CA, PIK3R5, TPR, TRAF6, FGF3, TRAF4, WNT8A, AXIN1, WNT8B, CEBPA, TCF7, MSH3, RELA, PIK3CD, ITGA3, MAPK10, DAPK2, STK4, FZD7, PIAS3, ARAF, VEGFA, PDGFRA, TCEB2, CRK*	1.60	1.95E-01
Focal adhesion	28	1.90	3.33E-03	*PDGFB, DIAPH1, PGF, ITGB3, VCL, PAK7, ITGB8, BCL2, TNR, RHOA, PIK3CA, PIK3R5, PAK1, COL11A2, FLT1, PIK3CD, ITGA3, MYL12A, MAPK10, FLNC, FLNB, KDR, VWF, ITGA5, VEGFA, PDGFRA, RELN, CRK*	1.78	1.76E-01
Cytokine-cytokine receptor interaction	33	2.24	6.47E-03	*TNF, PDGFB, CCR1, CSF1, CXCL9, IL13, TNFSF14, CNTFR, CX3CL1, CXCL11, ZFP91, IL10RB, EPO, FLT1, IL23R, TNFSF4, LTBR, EDA2R, IL6R, CD40, CCL16, KDR, CCR9, TNFRSF10A, OSM, IFNAR2, TNFRSF10D, VEGFA, PDGFRA, IL12B, NGFR, MPL, ACVR1*	1.61	2.46E-01
Inositol	11	0.75	7.85E-03	*PLCB3, PLCB4, INPP5K, INPP5J, PIK3CD, MIOX,*	2.61	2.40E-01

phosphate metabolism				*PIP5K1B, PLCD3, PIK3CA, PIP4K2C, INPP5A*		
Prostate cancer	15	1.02	8.26E-03	*E2F1, FGFR2, HSP90AB1, FGFR1, TCF7, PDGFB, RELA, PIK3CD, TCF7L2, BCL2, ARAF, PDGFRA, PIK3CA, PIK3R5, CREB3L3*	2.16	2.14E-01
Jak-STAT signalling pathway	22	1.49	8.31E-03	*IL23R, SOCS3, STAM2, PIK3CD, SOCS1, PIM1, IL13, CNTFR, IL6R, STAT2, TYK2, OSM, IFNAR2, ZFP91, PIAS3, IL10RB, PIK3CA, PIK3R5, STAM, IL12B, MPL, EPO*	1.82	1.87E-01
Acute myeloid leukemia	11	0.75	1.30E-02	*CEBPA, TCF7, EIF4EBP1, RELA, ARAF, PIK3CD, PIM1, PML, PIK3CA, PIK3R5, TCF7L2*	2.43	2.47E-01
MAPK signalling pathway	32	2.17	1.48E-02	*FGFR2, FGFR1, FGFR4, TNF, PDGFB, FGF14, MRAS, PPM1B, MAP3K7, DUSP16, PPP3CB, RRAS, PAK1, TRAF6, FGF3, RASA1, RELA, RELB, MAPK10, FLNC, CACNG1, STK4, FLNB, CDC25B, DUSP4, ARRB2, RPS6KA1, PDGFRA, CACNA1E, CRK, MAP3K12, MAP3K11*	1.54	2.51E-01
Renal cell carcinoma	12	0.81	1.83E-02	*PAK7, PDGFB, PGF, ARAF, PIK3CD, VEGFA, SLC2A1, TCEB2, PIK3CA, PIK3R5, PAK1, CRK*	2.20	2.75E-01
Axon guidance	18	1.22	2.14E-02	*PLXNB1, LIMK1, NTN4, L1CAM, EPHB3, EPHB1, PAK7, SEMA4G, SEMA6D, RGS3, CFL2, RHOA, SEMA4C, PPP3CB, SEMA4D, PAK1, UNC5C, RASA1*	1.79	2.90E-01
Toll-like receptor signalling pathway	15	1.02	2.37E-02	*TNF, RELA, PIK3CD, CXCL9, TIRAP, CD40, MAPK10, CXCL11, MAP3K7, IFNAR2, CASP8, PIK3CA, PIK3R5, IL12B, TRAF6*	1.90	2.94E-01
Phosphatidylinositol signalling system	12	0.81	2.68E-02	*PLCB3, PLCB4, INPP5K, INPP5J, PIK3CD, PIP5K1B, PLCD3, PIK3CA, PIK3R5, ITPR1, PIP4K2C, INPP5A*	2.08	3.05E-01
Endocytosis	23	1.56	2.74E-02	*FGFR2, PLD2, ARFGAP2, FGFR4, CLTA, FLT1, STAM2, PRKCI, PIP5K1B, KDR, DAB2, AP2B1, RABEP1, ARRB2, NEDD4, PDGFRA, VPS24, STAM, SMURF1, TRAF6, VPS28, SH3GL2, EPN2*	1.60	2.92E-01
Chemokine signalling pathway	23	1.56	3.22E-02	*RELA, CCR1, ADCY6, PIK3CD, CXCL9, FOXO3, CX3CL1, CCL16, CXCL11, STAT2, CCR9, PLCB3, GNGT2, DOCK2, PLCB4, ARRB2, RHOA, PIK3CA, PIK3R5, PAK1, GNB3, CSK, CRK*	1.58	3.16E-01
Apoptosis	13	0.88	3.59E-02	*IRAK2, TNF, DFFA, RELA, PIK3CD, TNFRSF10A, CASP6, TNFRSF10D, BCL2, CASP8, PPP3CB, PIK3CA, PIK3R5*	1.91	3.28E-01
Neurotrophin signalling pathway	16	1.08	5.67E-02	*IRAK2, YWHAZ, RELA, PIK3CD, FOXO3, MAPK10, RPS6KA1, BCL2, RHOA, PIK3CA, PIK3R5, NGFR, SH2B1, TRAF6, CSK, CRK*	1.65	4.50E-01
Colorectal cancer	12	0.81	5.97E-02	*TCF7, MSH3, BCL2, ARAF, PIK3CD, PDGFRA, PIK3CA, PIK3R5, MAPK10, TCF7L2, FZD7, AXIN1*	1.83	4.48E-01
Fc gamma R-mediated phagocytosis	13	0.88	6.35E-02	*PTPRC, PLD2, LIMK1, PIK3CD, PIP5K1B, ARPC4, DOCK2, GAB2, CFL2, PIK3CA, PIK3R5, PAK1, CRK*	1.75	4.52E-01
Type II diabetes mellitus	8	0.54	6.93E-02	*TNF, SOCS3, SOCS1, PIK3CD, PIK3CA, PIK3R5, CACNA1E, MAPK10*	2.18	4.64E-01
Melanoma	10	0.68	9.80E-02	*E2F1, FGFR1, PDGFB, FGF14, ARAF, PIK3CD, PDGFRA, PIK3CA, PIK3R5, FGF3*	1.80	5.74E-01

Table D.3 Pathway enrichment analysis results for predicted high-confidence triplexes.

References

Abeel, T., Helleputte, T., Van de Peer, Y., Dupont, P., and Saeys, Y. (2009). Robust biomarker identification for cancer diagnosis with ensemble feature selection methods. Bioinformatics *26*, 392–398.

Agirre, X., Vilas-Zornoza, A., Jiménez-Velasco, A., Martin-Subero, J.I., Cordeu, L., Gárate, L., San José-Eneriz, E., Abizanda, G., Rodríguez-Otero, P., Fortes, P., *et al.* (2009). Epigenetic silencing of the tumor suppressor microRNA Hsa-miR-124a regulates CDK6 expression and confers a poor prognosis in acute lymphoblastic leukemia. Cancer Res. *69*, 4443–4453.

Aguda, B.D. (2013). Modeling microRNA-transcription factor networks in cancer. Adv. Exp. Med. Biol. *774*, 149–167.

Aguda, B.D., Kim, Y., Piper-Hunter, M.G., Friedman, A., and Marsh, C.B. (2008). MicroRNA regulation of a cancer network: consequences of the feedback loops involving miR-17-92, E2F, and Myc. Proc Natl Acad Sci *105*, 19678–19683.

Albert, R., Jeong, H., and Barabási, A.-L. (2000). Error and attack tolerance of complex networks. Nature *406*, 378–382.

Alexiou, P., Vergoulis, T., Gleditzsch, M., Prekas, G., Dalamagas, T., Megraw, M., Grosse, I., Sellis, T., and Hatzigeorgiou, A.G. (2010). miRGen 2.0: a database of microRNA genomic information and regulation. Nucleic Acids Res. *38*, D137–D141.

Alla, V., Kowtharapu, B.S., Engelmann, D., Emmrich, S., Schmitz, U., Steder, M., and Pützer, B.M. (2012). E2F1 confers anticancer drug resistance by targeting ABC transporter family members and Bcl-2 via the p73/DNp73-miR-205 circuitry. Cell Cycle *11*, 3067–3078.

Alon, U. (2007a). Network motifs: theory and experimental approaches. Nat. Rev. Genet. *8*, 450–461.

Alon, U. (2007b). An introduction to systems biology: design principles of biological circuits (Boca Raton, FL: Chapman & Hall/CRC).

Altschul, S.F., Gish, W., Miller, W., Myers, E.W., and Lipman, D.J. (1990). Basic local alignment search tool. J. Mol. Biol. *215*, 403–410.

Alves, R., and Savageau, M.A. (2000). Extending the method of mathematically controlled comparison to include numerical comparisons. Bioinforma. Oxf. Engl. *16*, 786–798.

Ambros, V. (2004). The functions of animal microRNAs. Nature *431*, 350–355.

Ashburner, M., Ball, C.A., Blake, J.A., Botstein, D., Butler, H., Cherry, J.M., Davis, A.P., Dolinski, K., Dwight, S.S., Eppig, J.T., *et al.* (2000). Gene ontology: tool for the unification of biology. The Gene Ontology Consortium. Nat. Genet. *25*, 25–29.

Assenov, Y., Ramirez, F., Schelhorn, S.-E., Lengauer, T., and Albrecht, M. (2007). Computing topological parameters of biological networks. Bioinformatics *24*, 282–284.

Bader, J.S., Chaudhuri, A., Rothberg, J.M., and Chant, J. (2004). Gaining confidence in high-throughput protein interaction networks. Nat. Biotechnol. *22*, 78–85.

Balaga, O., Friedman, Y., and Linial, M. (2012). Toward a combinatorial nature of microRNA regulation in human cells. Nucleic Acids Res. *40*, 9404–9416.

Balsa-Canto, E., Alonso, A.A., and Banga, J.R. (2010). An iterative identification procedure for dynamic modeling of biochemical networks. BMC Syst. Biol. *4*, 11.

Banáš, P., Jurečka, P., Walter, N.G., Šponer, J., and Otyepka, M. (2009). Theoretical studies of RNA catalysis: Hybrid QM/MM methods and their comparison with MD and QM. Methods *49*, 202–216.

Bandyopadhyay, S., and Bhattacharyya, M. (2010). PuTmiR: a database for extracting neighboring transcription factors of human microRNAs. BMC Bioinformatics *11*, 190.

Bandyopadhyay, S., and Mitra, R. (2009). TargetMiner: microRNA target prediction with systematic identification of tissue-specific negative examples. Bioinformatics *25*, 2625–2631.

Barabási, A.-L., and Oltvai, Z.N. (2004). Network biology: understanding the cell's functional organization. Nat. Rev. Genet. *5*, 101–113.

Barbarotto, E., Schmittgen, T.D., and Calin, G.A. (2008). MicroRNAs and cancer: profile, profile, profile. Int J Cancer *122*, 969–977.

Bartel, D.P. (2004). MicroRNAs: genomics, biogenesis, mechanism, and function. Cell *116*, 281–297.

Bartel, D.P. (2009). MicroRNAs: target recognition and regulatory functions. Cell *136*, 215–233.

Le Béchec, A., Portales-Casamar, E., Vetter, G., Moes, M., Zindy, P.-J., Saumet, A., Arenillas, D., Theillet, C., Wasserman, W.W., Lecellier, C.-H., *et al.* (2011). MIR@NT@N: a framework integrating transcription factors, microRNAs and their targets to identify sub-network motifs in a meta-regulation network model. BMC Bioinformatics *12*, 67.

Berger, A.J., Davis, D.W., Tellez, C., Prieto, V.G., Gershenwald, J.E., Johnson, M.M., Rimm, D.L., and Bar-Eli, M. (2005). Automated quantitative analysis of activator protein-2alpha subcellular expression in melanoma tissue microarrays correlates with survival prediction. Cancer Res. *65*, 11185–11192.

Berman, H.M. (2000). The Protein Data Bank. Nucleic Acids Res. *28*, 235–242.

Betel, D., Wilson, M., Gabow, A., Marks, D.S., and Sander, C. (2008). The microRNA.org resource: targets and expression. Nucleic Acids Res *36*, D149–D153.

Betel, D., Koppal, A., Agius, P., Sander, C., and Leslie, C. (2010). Comprehensive modeling of microRNA targets predicts functional non-conserved and non-canonical sites. Genome Biol *11*, R90.

Bhattacharya, A., and Kunz, M. (2013). Target Identification, microRNA. In Encyclopedia of Systems Biology, W. Dubitzky, O. Wolkenhauer, K.-H. Cho, and H. Yokota, eds. (Springer New York), pp. 2138–2142.

Bhattacharya, A., Schmitz, U., Wolkenhauer, O., Schönherr, M., Raatz, Y., and Kunz, M. (2012). Regulation of cell cycle checkpoint kinase WEE1 by miR-195 in malignant melanoma. Oncogene 3175–3183.

Bhattacharya, A., Schmitz, U., Raatz, Y., Schönherr, M., Kottek, T., Schauer, M., Franz, S., Saalbach, A., Anderegg, U., Wolkenhauer, O., *et al.* (2015). miR-638 promotes melanoma metastasis and protects melanoma cells from apoptosis and autophagy. Oncotarget, accepted for publication.

Bhattacharyya, S.N., Habermacher, R., Martine, U., Closs, E.I., and Filipowicz, W. (2006). Relief of microRNA-mediated translational repression in human cells subjected to stress. Cell *125*, 1111–1124.

Borgdorff, V., Lleonart, M.E., Bishop, C.L., Fessart, D., Bergin, A.H., Overhoff, M.G., and Beach, D.H. (2010). Multiple microRNAs rescue from Ras-induced senescence by inhibiting p21(Waf1/Cip1). Oncogene *29*, 2262–2271.

Borneman, A.R., Leigh-Bell, J.A., Yu, H., Bertone, P., Gerstein, M., and Snyder, M. (2006). Target hub proteins serve as master regulators of development in yeast. Genes Dev. *20*, 435–448.

Brazma, A. (2003). ArrayExpress--a public repository for microarray gene expression data at the EBI. Nucleic Acids Res. *31*, 68–71.

Brennecke, J., Hipfner, D.R., Stark, A., Russell, R.B., and Cohen, S.M. (2003). bantam encodes a developmentally regulated microRNA that controls cell proliferation and regulates the proapoptotic gene hid in Drosophila. Cell *113*, 25–36.

Brigand, K.L., Robbe-Sermesant, K., Mari, B., and Barbry, P. (2010). MiRonTop: mining microRNAs targets across large scale gene expression studies. Bioinformatics *26*.

Bulyk, M.L. (2003). Computational prediction of transcription-factor binding site locations. Genome Biol. *5*, 201.

Bushati, N., and Cohen, S.M. (2007). microRNA functions. Annu Rev Cell Dev Biol *23*, 175–205.

Calin, G.A., and Croce, C.M. (2006). MicroRNA signatures in human cancers. Nat Rev Cancer *6*, 857–866.

Carrington, J.C. (2003). Role of MicroRNAs in Plant and Animal Development. Science *301*, 336–338.

Cascione, L., Ferro, A., Giugno, R., Laganà, A., Pigola, G., Pulvirenti, A., and Veneziano, D. (2013). Elucidating the role of microRNAs in cancer through data mining techniques. Adv. Exp. Med. Biol. *774*, 291–315.

Ceol, A., Chatr Aryamontri, A., Licata, L., Peluso, D., Briganti, L., Perfetto, L., Castagnoli, L., and Cesareni, G. (2010). MINT, the molecular interaction database: 2009 update. Nucleic Acids Res. *38*, D532–D539.

Chan, T.A., Hwang, P.M., Hermeking, H., Kinzler, K.W., and Vogelstein, B. (2000). Cooperative effects of genes controlling the G(2)/M checkpoint. Genes Dev. *14*, 1584–1588.

Chang, T.-C., and Mendell, J.T. (2007). microRNAs in Vertebrate Physiology and Human Disease. Annu. Rev. Genomics Hum. Genet. *8*, 215–239.

Chen, C.-Z. (2004). MicroRNAs Modulate Hematopoietic Lineage Differentiation. Science *303*, 83–86.

Chen, X., Ba, Y., Ma, L., Cai, X., Yin, Y., Wang, K., Guo, J., Zhang, Y., Chen, J., Guo, X., *et al.* (2008). Characterization of microRNAs in serum: a novel class of biomarkers for diagnosis of cancer and other diseases. Cell Res. *18*, 997–1006.

Chi, S.W., Hannon, G.J., and Darnell, R.B. (2012). An alternative mode of microRNA target recognition. Nat Struct Mol Biol *19*, 321–327.

Cho, S., Jang, I., Jun, Y., Yoon, S., Ko, M., Kwon, Y., Choi, I., Chang, H., Ryu, D., Lee, B., *et al.* (2012). miRGator v3.0: a microRNA portal for deep sequencing, expression profiling and mRNA targeting. Nucleic Acids Res. *41*, D252–D257.

Darty, K., Denise, A., and Ponty, Y. (2009). VARNA: Interactive drawing and editing of the RNA secondary structure. Bioinformatics *25*, 1974–1975.

Deane, C.M., Salwiński, Ł., Xenarios, I., and Eisenberg, D. (2002). Protein interactions: two methods for assessment of the reliability of high throughput observations. Mol. Cell. Proteomics MCP *1*, 349–356.

el-Deiry, W.S., Harper, J.W., O'Connor, P.M., Velculescu, V.E., Canman, C.E., Jackman, J., Pietenpol, J.A., Burrell, M., Hill, D.E., and Wang, Y. (1994). WAF1/CIP1 is induced in p53-mediated G1 arrest and apoptosis. Cancer Res. *54*, 1169–1174.

Didiano, D., and Hobert, O. (2008). Molecular architecture of a miRNA-regulated 3' UTR. RNA *14*, 1297–1317.

Dinger, M.E., Pang, K.C., Mercer, T.R., and Mattick, J.S. (2008). Differentiating protein-coding and noncoding RNA: challenges and ambiguities. PLoS Comput Biol *4*, e1000176.

Dirks, R.M., Bois, J.S., Schaeffer, J.M., Winfree, E., and Pierce, N.A. (2007). Thermodynamic Analysis of Interacting Nucleic Acid Strands. SIAM Rev. *49*, 65–88.

Doench, J.G., and Sharp, P.A. (2004). Specificity of microRNA target selection in translational repression. Genes Dev *18*, 504–511.

Duursma, A.M., Kedde, M., Schrier, M., le Sage, C., and Agami, R. (2008). miR-148 targets human DNMT3b protein coding region. RNA *14*, 872–877.

Dweep, H., Sticht, C., Pandey, P., and Gretz, N. (2011). miRWalk - Database: Prediction of possible miRNA binding sites by "walking" the genes of three genomes. J. Biomed. Inform. *44*, 839–847.

Eddy, S.R. (2004). How do RNA folding algorithms work? Nat. Biotechnol. *22*, 1457–1458.

Bar-Eli, M. (2001). Gene Regulation in Melanoma Progression by the AP-2 Transcription Factor. Pigment Cell Res. *14*, 78–85.

Elkayam, E., Kuhn, C.-D., Tocilj, A., Haase, A.D., Greene, E.M., Hannon, G.J., and Joshua-Tor, L. (2012). The Structure of Human Argonaute-2 in Complex with miR-20a. Cell *150*, 100–110.

El-Metwally, S., Hamza, T., Zakaria, M., and Helmy, M. (2013). Next-Generation Sequence Assembly: Four Stages of Data Processing and Computational Challenges. PLoS Comput. Biol. *9*, e1003345.

Engels, B.M., and Hutvagner, G. (2006). Principles and effects of microRNA-mediated post-transcriptional gene regulation. Oncogene *25*, 6163–6169.

Enright, A.J., John, B., Gaul, U., Tuschl, T., Sander, C., and Marks, D.S. (2003). MicroRNA targets in Drosophila. Genome Biol *5*, R1.

Erler, J.T., and Linding, R. (2009). Network-based drugs and biomarkers. J. Pathol. 290–296.

Fan, X., and Kurgan, L. (2014). Comprehensive overview and assessment of computational prediction of microRNA targets in animals. Brief. Bioinform.

Farazi, T.A., Hoell, J.I., Morozov, P., and Tuschl, T. (2013). MicroRNAs in Human Cancer. In MicroRNA Cancer Regulation, U. Schmitz, O. Wolkenhauer, and J. Vera, eds. (Dordrecht: Springer Netherlands), pp. 1–20.

Farh, K.K.-H., Grimson, A., Jan, C., Lewis, B.P., Johnston, W.K., Lim, L.P., Burge, C.B., and Bartel, D.P. (2005). The widespread impact of mammalian MicroRNAs on mRNA repression and evolution. Science *310*, 1817–1821.

Fazekas, D., Koltai, M., Türei, D., Módos, D., Pálfy, M., Dúl, Z., Zsákai, L., Szalay-Bekő, M., Lenti, K., Farkas, I.J., *et al.* (2013). SignaLink 2 – a signaling pathway resource with multi-layered regulatory networks. BMC Syst. Biol. *7*, 7.

Ferracin, M., Veronese, A., and Negrini, M. (2010). Micromarkers: miRNAs in cancer diagnosis and prognosis. Expert Rev Mol Diagn *10*, 297–308.

Filipowicz, W., Bhattacharyya, S.N., and Sonenberg, N. (2008). Mechanisms of post-transcriptional regulation by microRNAs: are the answers in sight? Nat Rev Genet *9*, 102–114.

Foloppe, N., and MacKerell, Jr., A.D. (2000). All-atom empirical force field for nucleic acids: I. Parameter optimization based on small molecule and condensed phase macromolecular target data. J. Comput. Chem. *21*, 86–104.

Freiesleben, S., Schmitz, U., and Vera, J. (2013). MicroRNA, Biogenesis, Regulation. In Encyclopedia of Systems Biology, W. Dubitzky, O. Wolkenhauer, K.-H. Cho, and H. Yokota, eds. (Springer New York), pp. 1306–1310.

Friedländer, M.R., Chen, W., Adamidi, C., Maaskola, J., Einspanier, R., Knespel, S., and Rajewsky, N. (2008). Discovering microRNAs from deep sequencing data using miRDeep. Nat. Biotechnol. *26*, 407–415.

Friedman, R.C., Farh, K.K.-H., Burge, C.B., and Bartel, D.P. (2009). Most mammalian mRNAs are conserved targets of microRNAs. Genome Res *19*, 92–105.

Friedman, Y., Balaga, O., and Linial, M. (2013). Working Together: Combinatorial Regulation by microRNAs. In MicroRNA Cancer Regulation, U. Schmitz, O. Wolkenhauer, and J. Vera, eds. (Dordrecht: Springer Netherlands), pp. 317–337.

Funahashi, A., Morohashi, M., Kitano, H., and Tanimura, N. (2003). CellDesigner: a process diagram editor for gene-regulatory and biochemical networks. BIOSILICO *1*, 159–162.

Furuta, M., Kozaki, K., Tanaka, S., Arii, S., Imoto, I., and Inazawa, J. (2010). miR-124 and miR-203 are epigenetically silenced tumor-suppressive microRNAs in hepatocellular carcinoma. Carcinogenesis *31*, 766–776.

Gabriely, G., Wurdinger, T., Kesari, S., Esau, C.C., Burchard, J., Linsley, P.S., and Krichevsky, A.M. (2008). MicroRNA 21 Promotes Glioma Invasion by Targeting Matrix Metalloproteinase Regulators. Mol. Cell. Biol. *28*, 5369–5380.

Gan, H.H., and Gunsalus, K.C. (2013). Tertiary structure-based analysis of microRNA-target interactions. RNA *19*, 539–551.

Gardner, P.P., and Giegerich, R. (2004). A comprehensive comparison of comparative RNA structure prediction approaches. BMC Bioinformatics *5*, 140.

Gartel, A.L., and Tyner, A.L. (1999). Transcriptional regulation of the p21((WAF1/CIP1)) gene. Exp. Cell Res. *246*, 280–289.

Garzon, R., Calin, G.A., and Croce, C.M. (2009). MicroRNAs in Cancer. Annu Rev Med *60*, 167–179.

Gilbert, W. (1986). Origin of life: The RNA world. Nature *319*, 618–618.

Griffiths-Jones, S. (2004). The microRNA Registry. Nucleic Acids Res. *32*, D109–D111.

Griffiths-Jones, S., Grocock, R.J., Dongen, S. van, Bateman, A., and Enright, A.J. (2006). miRBase: microRNA sequences, targets and gene nomenclature. Nucleic Acids Res *34*, D140–D144.

Grimson, A., Farh, K.K.-H., Johnston, W.K., Garrett-Engele, P., Lim, L.P., and Bartel, D.P. (2007). MicroRNA targeting specificity in mammals: determinants beyond seed pairing. Mol. Cell *27*, 91–105.

Gupta, S.K., and Schmitz, U. (2011). Bioinformatics Analysis of High-Throughput Experiments. In Recent Trends in Biotechnology, M.P. Singh, A. Agrawal, and B. Sharma, eds. (Nova Science Publishers, Inc.), pp. 129–156.

Hafner, M., Landthaler, M., Burger, L., Khorshid, M., Hausser, J., Berninger, P., Rothballer, A., Ascano, M., Jungkamp, A.-C., Munschauer, M., *et al.* (2010). Transcriptome-wide identification of RNA-binding protein and microRNA target sites by PAR-CLIP. Cell *141*, 129–141.

Hammell, M., Long, D., Zhang, L., Lee, A., Carmack, C.S., Han, M., Ding, Y., and Ambros, V. (2008). mirWIP: microRNA target prediction based on microRNA-containing ribonucleoprotein-enriched transcripts. Nat Methods *5*, 813–819.

Harper, J.W., Adami, G.R., Wei, N., Keyomarsi, K., and Elledge, S.J. (1993). The p21 Cdk-interacting protein Cip1 is a potent inhibitor of G1 cyclin-dependent kinases. Cell *75*, 805–816.

Hart, K., Nyström, B., Ohman, M., and Nilsson, L. (2005). Molecular dynamics simulations and free energy calculations of base flipping in dsRNA. RNA *11*, 609–618.

He, L., and Hannon, G.J. (2004). MicroRNAs: small RNAs with a big role in gene regulation. Nat Rev Genet *5*, 522–531.

Hecker, M., Lambeck, S., Toepfer, S., van Someren, E., and Guthke, R. (2009). Gene regulatory network inference: Data integration in dynamic models—A review. Biosystems *96*, 86–103.

Helwak, A., Kudla, G., Dudnakova, T., and Tollervey, D. (2013). Mapping the human miRNA interactome by CLASH reveals frequent noncanonical binding. Cell *153*, 654–665.

Hendrickson, D.G., Hogan, D.J., McCullough, H.L., Myers, J.W., Herschlag, D., Ferrell, J.E., and Brown, P.O. (2009). Concordant Regulation of Translation and mRNA Abundance for Hundreds of Targets of a Human microRNA. PLoS Biol. *7*, e1000238.

Herbig, A., and Nieselt, K. (2013). non-coding RNA prediction. In Encyclopedia of Systems Biology, W. Dubitzky, O. Wolkenhauer, K.-H. Cho, and H. Yokota, eds. (Springer New York), pp. 1534–1538.

Herranz, H., and Cohen, S.M. (2010). MicroRNAs and gene regulatory networks: managing the impact of noise in biological systems. Genes Dev. *24*, 1339–1344.

Hill, C.G., Matyunina, L.V., Walker, D., Benigno, B.B., and McDonald, J.F. (2014). Transcriptional override: a regulatory network model of indirect responses to modulations in microRNA expression. BMC Syst. Biol. *8*, 36.

Hofacker, I.L. (2003). Vienna RNA secondary structure server. Nucleic Acids Res. *31*, 3429–3431.

Hofacker, I.L. (2009). RNA secondary structure analysis using the Vienna RNA package. Curr. Protoc. Bioinforma. Ed. Board Andreas Baxevanis Al *Chapter 12*, Unit12.2.

Hon, L.S., and Zhang, Z. (2007). The roles of binding site arrangement and combinatorial targeting in microRNA repression of gene expression. Genome Biol. *8*, R166.

Hoops, S., Sahle, S., Gauges, R., Lee, C., Pahle, J., Simus, N., Singhal, M., Xu, L., Mendes, P., and Kummer, U. (2006). COPASI--a COmplex PAthway SImulator. Bioinformatics *22*, 3067–3074.

Hsu, P.W.C., Huang, H.-D., Hsu, S.-D., Lin, L.-Z., Tsou, A.-P., Tseng, C.-P., Stadler, P.F., Washietl, S., and Hofacker, I.L. (2006). miRNAMap: genomic maps of microRNA genes and their target genes in mammalian genomes. Nucleic Acids Res *34*, D135–D139.

Hsu, S.-D., Chu, C.-H., Tsou, A.-P., Chen, S.-J., Chen, H.-C., Hsu, P.W.-C., Wong, Y.-H., Chen, Y.-H., Chen, G.-H., and Huang, H.-D. (2008). miRNAMap 2.0: genomic maps of microRNAs in metazoan genomes. Nucleic Acids Res. *36*, D165–D169.

Hsu, S.-D., Lin, F.-M., Wu, W.-Y., Liang, C., Huang, W.-C., Chan, W.-L., Tsai, W.-T., Chen, G.-Z., Lee, C.-J., Chiu, C.-M., *et al.* (2011). miRTarBase: a database curates experimentally validated microRNA-target interactions. Nucleic Acids Res. *39*, D163–D169.

Hsu, S.-D., Tseng, Y.-T., Shrestha, S., Lin, Y.-L., Khaleel, A., Chou, C.-H., Chu, C.-F., Huang, H.-Y., Lin, C.-M., Ho, S.-Y., *et al.* (2013). miRTarBase update 2014: an information resource for experimentally validated miRNA-target interactions. Nucleic Acids Res.

Huang, D.W., Sherman, B.T., and Lempicki, R.A. (2008). Bioinformatics enrichment tools: paths toward the comprehensive functional analysis of large gene lists. Nucleic Acids Res. *37*, 1–13.

Huang, D.W., Sherman, B.T., and Lempicki, R.A. (2009). Systematic and integrative analysis of large gene lists using DAVID bioinformatics resources. Nat. Protoc. *4*, 44–57.

Huang, J.C., Babak, T., Corson, T.W., Chua, G., Khan, S., Gallie, B.L., Hughes, T.R., Blencowe, B.J., Frey, B.J., and Morris, Q.D. (2007). Using expression profiling data to identify human microRNA targets. Nat. Methods *4*, 1045–1049.

Hucka, M., Finney, A., Sauro, H.M., Bolouri, H., Doyle, J.C., Kitano, H., and the rest of the SBML Forum:, Arkin, A.P., Bornstein, B.J., Bray, D., *et al.* (2003). The systems biology markup language (SBML): a medium for representation and exchange of biochemical network models. Bioinformatics *19*, 524–531.

Hüttenhofer, A., and Vogel, J. (2006). Experimental approaches to identify non-coding RNAs. Nucleic Acids Res *34*, 635–646.

Hutvágner, G., and Zamore, P.D. (2002). RNAi: nature abhors a double-strand. Curr. Opin. Genet. Dev. *12*, 225–232.

Hwang, H.-W., Wentzel, E.A., and Mendell, J.T. (2007). A hexanucleotide element directs microRNA nuclear import. Science *315*, 97–100.

Ideker, T., Ozier, O., Schwikowski, B., and Siegel, A.F. (2002). Discovering regulatory and signalling circuits in molecular interaction networks. Bioinforma. Oxf. Engl. *18 Suppl 1*, S233–S240.

Im, W., Lee, M.S., and Brooks, C.L., 3rd (2003). Generalized born model with a simple smoothing function. J. Comput. Chem. *24*, 1691–1702.

Jeffares, D.C., Poole, A.M., and Penny, D. (1998). Relics from the RNA world. J. Mol. Evol. *46*, 18–36.

Jiang, L., Huang, Q., Zhang, S., Zhang, Q., Chang, J., Qiu, X., and Wang, E. (2010a). Hsa-miR-125a-3p and hsa-miR-125a-5p are downregulated in non-small cell lung cancer and have inverse effects on invasion and migration of lung cancer cells. BMC Cancer *10*, 318.

Jiang, Q., Feng, M.-G., and Mo, Y.-Y. (2009a). Systematic validation of predicted microRNAs for cyclin D1. BMC Cancer *9*, 194.

Jiang, Q., Wang, Y., Hao, Y., Juan, L., Teng, M., Zhang, X., Li, M., Wang, G., and Liu, Y. (2009b). miR2Disease: a manually curated database for microRNA deregulation in human disease. Nucleic Acids Res *37*, D98–D104.

Jiang, Q., Hao, Y., Wang, G., Juan, L., Zhang, T., Teng, M., Liu, Y., and Wang, Y. (2010b). Prioritization of disease microRNAs through a human phenome-microRNAome network. BMC Syst Biol *4 Suppl 1*, S2.

John, B., Enright, A.J., Aravin, A., Tuschl, T., Sander, C., and Marks, D.S. (2004). Human MicroRNA targets. PLoS Biol *2*, e363.

Johnston, R.J., Chang, S., Etchberger, J.F., Ortiz, C.O., and Hobert, O. (2005). MicroRNAs acting in a double-negative feedback loop to control a neuronal cell fate decision. Proc. Natl. Acad. Sci. *102*, 12449–12454.

Jones-Rhoades, M.W., Bartel, D.P., and Bartel, B. (2006). MicroRNAS and their regulatory roles in plants. Annu. Rev. Plant Biol. *57*, 19–53.

Jung, Y.-S., Qian, Y., and Chen, X. (2010). Examination of the expanding pathways for the regulation of p21 expression and activity. Cell. Signal. *22*, 1003–1012.

Kai, Z.S., and Pasquinelli, A.E. (2010). MicroRNA assassins: factors that regulate the disappearance of miRNAs. Nat Struct Mol Biol *17*, 5–10.

Kamburov, A., Pentchev, K., Galicka, H., Wierling, C., Lehrach, H., and Herwig, R. (2011). ConsensusPathDB: toward a more complete picture of cell biology. Nucleic Acids Res. *39*, D712–D717.

Kanehisa, M., Araki, M., Goto, S., Hattori, M., Hirakawa, M., Itoh, M., Katayama, T., Kawashima, S., Okuda, S., Tokimatsu, T., *et al.* (2008). KEGG for linking genomes to life and the environment. Nucleic Acids Res. *36*, D480–D484.

Karolchik, D., Baertsch, R., Diekhans, M., Furey, T.S., Hinrichs, A., Lu, Y.T., Roskin, K.M., Schwartz, M., Sugnet, C.W., Thomas, D.J., *et al.* (2003). The UCSC Genome Browser Database. Nucleic Acids Res. *31*, 51–54.

Kasinski, A.L., and Slack, F.J. (2011). MicroRNAs en route to the clinic: progress in validating and targeting microRNAs for cancer therapy. Nat. Rev. Cancer *11*, 849–864.

Kerrien, S., Aranda, B., Breuza, L., Bridge, A., Broackes-Carter, F., Chen, C., Duesbury, M., Dumousseau, M., Feuermann, M., Hinz, U., *et al.* (2012). The IntAct molecular interaction database in 2012. Nucleic Acids Res. *40*, D841–D846.

Kertesz, M., Iovino, N., Unnerstall, U., Gaul, U., and Segal, E. (2007). The role of site accessibility in microRNA target recognition. Nat Genet *39*, 1278–1284.

Keshava Prasad, T.S., Goel, R., Kandasamy, K., Keerthikumar, S., Kumar, S., Mathivanan, S., Telikicherla, D., Raju, R., Shafreen, B., Venugopal, A., *et al.* (2009). Human Protein Reference Database–2009 update. Nucleic Acids Res. *37*, D767–D772.

Khanin, R., and Vinciotti, V. (2008). Computational modeling of post-transcriptional gene regulation by microRNAs. J Comput Biol *15*, 305–316.

Knoll, S., Furst, K., Kowtharapu, B., Schmitz, U., Marquardt, S., Wolkenhauer, O., Martin, H., and Putzer, B.M. (2014b). E2F1 induces miR-224/452 expression to drive EMT through TXNIP downregulation. EMBO Rep. *15*, 1315–1329.

Kozomara, A., and Griffiths-Jones, S. (2011). miRBase: integrating microRNA annotation and deep-sequencing data. Nucleic Acids Res *39*, D152–D157.

Kozomara, A., and Griffiths-Jones, S. (2014). miRBase: annotating high confidence microRNAs using deep sequencing data. Nucleic Acids Res. *42*, D68–D73.

Krek, A., Grün, D., Poy, M.N., Wolf, R., Rosenberg, L., Epstein, E.J., MacMenamin, P., Piedade, I. da, Gunsalus, K.C., Stoffel, M., *et al.* (2005). Combinatorial microRNA target predictions. Nat Genet *37*, 495–500.

Krol, J., Loedige, I., and Filipowicz, W. (2010). The widespread regulation of microRNA biogenesis, function and decay. Nat. Rev. Genet. *11*, 597–610.

Kudla, G., Granneman, S., Hahn, D., Beggs, J.D., and Tollervey, D. (2011). Cross-linking, ligation, and sequencing of hybrids reveals RNA-RNA interactions in yeast. Proc. Natl. Acad. Sci. *108*, 10010–10015.

Laganà, A., Forte, S., Giudice, A., Arena, M.R., Puglisi, P.L., Giugno, R., Pulvirenti, A., Shasha, D., and Ferro, A. (2009). miRò: a miRNA knowledge base. Database Oxf. *2009*, bap008.

Lagos-Quintana, M., Rauhut, R., Lendeckel, W., and Tuschl, T. (2001). Identification of novel genes coding for small expressed RNAs. Science *294*, 853–858.

Lai, X., and Vera, J. (2013). MicroRNA clusters. In Encyclopedia of Systems Biology, W. Dubitzky, O. Wolkenhauer, K.-H. Cho, and H. Yokota, eds. (Springer New York), pp. 1310–1314.

Lai, X., Schmitz, U., Gupta, S.K., Bhattacharya, A., Kunz, M., Wolkenhauer, O., and Vera, J. (2012). Computational analysis of target hub gene repression regulated by multiple and cooperative miRNAs. Nucleic Acids Res. *40*, 8818–8834.

Lai, X., Bhattacharya, A., Schmitz, U., Kunz, M., Vera, J., and Wolkenhauer, O. (2013). A Systems Biology Approach to Study MicroRNA-Mediated Gene Regulatory Networks. BioMed Res. Int. *2013*, 1–15.

Landgraf, P., Rusu, M., Sheridan, R., Sewer, A., Iovino, N., Aravin, A., Pfeffer, S., Rice, A., Kamphorst, A.O., Landthaler, M., *et al.* (2007). A mammalian microRNA expression atlas based on small RNA library sequencing. Cell *129*, 1401–1414.

Larkin, M.A., Blackshields, G., Brown, N.P., Chenna, R., McGettigan, P.A., McWilliam, H., Valentin, F., Wallace, I.M., Wilm, A., Lopez, R., *et al.* (2007). Clustal W and Clustal X version 2.0. Bioinforma. Oxf. Engl. *23*, 2947–2948.

Lau, N.C., Lim, L.P., Weinstein, E.G., and Bartel, D.P. (2001). An abundant class of tiny RNAs with probable regulatory roles in Caenorhabditis elegans. Science *294*, 858–862.

Lee, R.C., and Ambros, V. (2001). An extensive class of small RNAs in Caenorhabditis elegans. Science *294*, 862–864.

Lee, R.C., Feinbaum, R.L., and Ambros, V. (1993). The C. elegans heterochronic gene lin-4 encodes small RNAs with antisense complementarity to lin-14. Cell *75*, 843–854.

Lee, Y., Yang, X., Huang, Y., Fan, H., Zhang, Q., Wu, Y., Li, J., Hasina, R., Cheng, C., Lingen, M.W., *et al.* (2010). Network modeling identifies molecular functions targeted by miR-204 to suppress head and neck tumor metastasis. PLoS Comput Biol *6*, e1000730.

Levine, E., Jacob, E.B., and Levine, H. (2007). Target-specific and global effectors in gene regulation by MicroRNA. Biophys J *93*, L52–L54.

Lewis, B.P., Shih, I., Jones-Rhoades, M.W., Bartel, D.P., and Burge, C.B. (2003). Prediction of mammalian microRNA targets. Cell *115*, 787–798.

Lewis, B.P., Burge, C.B., and Bartel, D.P. (2005). Conserved seed pairing, often flanked by adenosines, indicates that thousands of human genes are microRNA targets. Cell *120*, 15–20.

Li, C., Donizelli, M., Rodriguez, N., Dharuri, H., Endler, L., Chelliah, V., Li, L., He, E., Henry, A., Stefan, M.I., *et al.* (2010). BioModels Database: An enhanced, curated and annotated resource for published quantitative kinetic models. BMC Syst. Biol. *4*, 92.

Li, D., Liu, W., Liu, Z., Wang, J., Liu, Q., Zhu, Y., and He, F. (2008). PRINCESS, a Protein Interaction Confidence Evaluation System with Multiple Data Sources. Mol. Cell. Proteomics *7*, 1043–1052.

Li, Y., Qiu, C., Tu, J., Geng, B., Yang, J., Jiang, T., and Cui, Q. (2014). HMDD v2.0: a database for experimentally supported human microRNA and disease associations. Nucleic Acids Res. *42*, D1070–D1074.

Licatalosi, D.D., Mele, A., Fak, J.J., Ule, J., Kayikci, M., Chi, S.W., Clark, T.A., Schweitzer, A.C., Blume, J.E., Wang, X., *et al.* (2008). HITS-CLIP yields genome-wide insights into brain alternative RNA processing. Nature *456*, 464–469.

Lim, L.P., Lau, N.C., Garrett-Engele, P., Grimson, A., Schelter, J.M., Castle, J., Bartel, D.P., Linsley, P.S., and Johnson, J.M. (2005). Microarray analysis shows that some microRNAs downregulate large numbers of target mRNAs. Nature *433*, 769–773.

Lorenz, R., Bernhart, S.H., Höner zu Siederdissen, C., Tafer, H., Flamm, C., Stadler, P.F., and Hofacker, I.L. (2011). ViennaRNA Package 2.0. Algorithms Mol. Biol. *6*, 26.

Lyngsø, R.B., and Pedersen, C.N.S. (2000). RNA Pseudoknot Prediction in Energy-Based Models. J. Comput. Biol. *7*, 409–427.

Lytle, J.R., Yario, T.A., and Steitz, J.A. (2007). Target mRNAs are repressed as efficiently by microRNA-binding sites in the 5' UTR as in the 3' UTR. Proc. Natl. Acad. Sci. U. S. A. *104*, 9667–9672.

Maki, C.G., and Howley, P.M. (1997). Ubiquitination of p53 and p21 is differentially affected by ionizing and UV radiation. Mol. Cell. Biol. *17*, 355–363.

Mangan, S., and Alon, U. (2003). Structure and function of the feed-forward loop network motif. Proc. Natl. Acad. Sci. U. S. A. *100*, 11980–11985.

Martinez, N.J., and Walhout, A.J.M. (2009). The interplay between transcription factors and microRNAs in genome-scale regulatory networks. BioEssays *31*, 435–445.

Martinez-Sanchez, A., and Murphy, C. (2013). MicroRNA Target Identification—Experimental Approaches. Biology *2*, 189–205.

McDowell, S.E., Špačková, N., Šponer, J., and Walter, N.G. (2007). Molecular dynamics simulations of RNA: Anin silico single molecule approach. Biopolymers *85*, 169–184.

Meister, G. (2013). Argonaute proteins: functional insights and emerging roles. Nat. Rev. Genet. *14*, 447–459.

Mendes, N. (2013). MicroRNA, Gene Prediction. In Encyclopedia of Systems Biology, W. Dubitzky, O. Wolkenhauer, K.-H. Cho, and H. Yokota, eds. (Springer New York), pp. 1341–1343.

Mendes, N.D., Freitas, A.T., and Sagot, M.-F. (2009). Current tools for the identification of miRNA genes and their targets. Nucleic Acids Res *37*, 2419–2433.

Mestdagh, P., Vlierberghe, P.V., Weer, A.D., Muth, D., Westermann, F., Speleman, F., and Vandesompele, J. (2009). A novel and universal method for microRNA RT-qPCR data normalization. Genome Biol *10*, R64.

Meyer, I.M. (2007). A practical guide to the art of RNA gene prediction. Brief. Bioinform. *8*, 396–414.

Meyer, L.R., Zweig, A.S., Hinrichs, A.S., Karolchik, D., Kuhn, R.M., Wong, M., Sloan, C.A., Rosenbloom, K.R., Roe, G., Rhead, B., *et al.* (2012). The UCSC Genome Browser database: extensions and updates 2013. Nucleic Acids Res. *41*, D64–D69.

Miranda, K.C., Huynh, T., Tay, Y., Ang, Y.-S., Tam, W.-L., Thomson, A.M., Lim, B., and Rigoutsos, I. (2006). A pattern-based method for the identification of MicroRNA binding sites and their corresponding heteroduplexes. Cell *126*, 1203–1217.

Mitchell, P.S., Parkin, R.K., Kroh, E.M., Fritz, B.R., Wyman, S.K., Pogosova-Agadjanyan, E.L., Peterson, A., Noteboom, J., O'Briant, K.C., Allen, A., *et al.* (2008). Circulating microRNAs as stable blood-based markers for cancer detection. Proc. Natl. Acad. Sci. *105*, 10513–10518.

Miyoshi, K., Miyoshi, T., and Siomi, H. (2010). Many ways to generate microRNA-like small RNAs: non-canonical pathways for microRNA production. Mol Genet Genomics *284*, 95–103.

Muckstein, U., Tafer, H., Hackermuller, J., Bernhart, S.H., Stadler, P.F., and Hofacker, I.L. (2006). Thermodynamics of RNA-RNA binding. Bioinformatics *22*, 1177–1182.

Mukherji, S., Ebert, M.S., Zheng, G.X.Y., Tsang, J.S., Sharp, P.A., and van Oudenaarden, A. (2011). MicroRNAs can generate thresholds in target gene expression. Nat. Genet. *43*, 854–859.

Nam, S., Kim, B., Shin, S., and Lee, S. (2008). miRGator: an integrated system for functional annotation of microRNAs. Nucleic Acids Res *36*, D159–D164.

Nikolov, S., and Vera, J. (2013). MicroRNA Regulation, Feedback Loop. In Encyclopedia of Systems Biology, W. Dubitzky, O. Wolkenhauer, K.-H. Cho, and H. Yokota, eds. (Springer New York), pp. 1320–1324.

Nikolov, S., Vera, J., Schmitz, U., and Wolkenhauer, O. (2010). A model-based strategy to investigate the role of microRNA regulation in cancer signalling networks. Theory Biosci. *130*, 55–69.

Nina, M., Beglov, D., and Roux, B. (1997). Atomic Radii for Continuum Electrostatics Calculations Based on Molecular Dynamics Free Energy Simulations. J. Phys. Chem. B *101*, 5239–5248.

Nissan, T., and Parker, R. (2008). Computational analysis of miRNA-mediated repression of translation: implications for models of translation initiation inhibition. RNA *14*, 1480–1491.

Le Novère, N., Hucka, M., Mi, H., Moodie, S., Schreiber, F., Sorokin, A., Demir, E., Wegner, K., Aladjem, M.I., Wimalaratne, S.M., *et al.* (2009). The Systems Biology Graphical Notation. Nat. Biotechnol. *27*, 735–741.

Osella, M., Bosia, C., Corá, D., and Caselle, M. (2011). The role of incoherent microRNA-mediated feedforward loops in noise buffering. PLoS Comput Biol *7*, e1001101.

Paciello, G., Acquaviva, A., Ficarra, E., Deriu, M.A., and Macii, E. (2011). A molecular dynamics study of a miRNA:mRNA interaction. J. Mol. Model. *17*, 2895–2906.

Paliy, M., Melnik, R., and Shapiro, B.A. (2009). Molecular dynamics study of the RNA ring nanostructure: a phenomenon of self-stabilization. Phys. Biol. *6*, 046003.

Papadopoulos, G.L., Reczko, M., Simossis, V.A., Sethupathy, P., and Hatzigeorgiou, A.G. (2009). The database of experimentally supported targets: a functional update of TarBase. Nucleic Acids Res *37*, D155–D158.

Parisien, M., and Major, F. (2008). The MC-Fold and MC-Sym pipeline infers RNA structure from sequence data. Nature *452*, 51–55.

Peng, Q., and Schork, N.J. (2014). Utility of network integrity methods in therapeutic target identification. Front. Genet. *5*.

Poole, A.M., Jeffares, D.C., and Penny, D. (1998). The path from the RNA world. J. Mol. Evol. *46*, 1–17.

Popenda, M., Szachniuk, M., Blazewicz, M., Wasik, S., Burke, E.K., Blazewicz, J., and Adamiak, R.W. (2010). RNA FRABASE 2.0: an advanced web-accessible database with the capacity to search the three-dimensional fragments within RNA structures. BMC Bioinformatics *11*, 231.

Popenda, M., Szachniuk, M., Antczak, M., Purzycka, K.J., Lukasiak, P., Bartol, N., Blazewicz, J., and Adamiak, R.W. (2012). Automated 3D structure composition for large RNAs. Nucleic Acids Res. *40*, e112–e112.

Raasch, P., Schmitz, U., Patenge, N., Vera, J., Kreikemeyer, B., and Wolkenhauer, O. (2010). Non-coding RNA detection methods combined to improve usability, reproducibility and precision. BMC Bioinformatics *11*, 491.

Rajewsky, N. (2006). microRNA target predictions in animals. Nat Genet *38*, S8–S13.

Rajewsky, N., and Socci, N.D. (2004). Computational identification of microRNA targets. Dev Biol *267*, 529–535.

Re, A., Corá, D., Taverna, D., and Caselle, M. (2009). Genome-wide survey of microRNA-transcription factor feed-forward regulatory circuits in human. Mol. Biosyst. *5*, 854–867.

Rehmsmeier, M., Steffen, P., Hochsmann, M., and Giegerich, R. (2004). Fast and effective prediction of microRNA/target duplexes. RNA *10*, 1507–1517.

Rehwinkel, J., Behm-Ansmant, I., Gatfield, D., and Izaurralde, E. (2005). A crucial role for GW182 and the DCP1:DCP2 decapping complex in miRNA-mediated gene silencing. RNA N. Y. N *11*, 1640–1647.

Ren, X., Wang, Y., Wang, J., and Zhang, X.-S. (2012). A unified computational model for revealing and predicting subtle subtypes of cancers. BMC Bioinformatics *13*, 70.

Rhoades, M.W., Reinhart, B.J., Lim, L.P., Burge, C.B., Bartel, B., and Bartel, D.P. (2002). Prediction of Plant MicroRNA Targets. Cell *110*, 513–520.

Rigoutsos, I. (2006). Short blocks from the noncoding parts of the human genome have instances within nearly all known genes and relate to biological processes. Proc. Natl. Acad. Sci. *103*, 6605–6610.

Riley, T., Sontag, E., Chen, P., and Levine, A. (2008). Transcriptional control of human p53-regulated genes. Nat. Rev. Mol. Cell Biol. *9*, 402–412.

Ritchie, W., Flamant, S., and Rasko, J.E.J. (2009). Predicting microRNA targets and functions: traps for the unwary. Nat Methods *6*, 397–398.

Ritchie, W., Rasko, J.E.J., and Flamant, S. (2013). MicroRNA target prediction and validation. Adv. Exp. Med. Biol. *774*, 39–53.

Rivas, E., and Eddy, S.R. (2000). Secondary structure alone is generally not statistically significant for the detection of noncoding RNAs. Bioinformatics *16*, 583–605.

Robins, H., Li, Y., and Padgett, R.W. (2005). Incorporating structure to predict microRNA targets. Proc Natl Acad Sci U A *102*, 4006–4009.

Rother, K., Rother, M., Boniecki, M., Puton, T., and Bujnicki, J.M. (2011). RNA and protein 3D structure modeling: similarities and differences. J. Mol. Model. *17*, 2325–2336.

Ruby, J.G., Jan, C.H., and Bartel, D.P. (2007). Intronic microRNA precursors that bypass Drosha processing. Nature *448*, 83–86.

Saetrom, P., Heale, B.S.E., Snøve, O., Aagaard, L., Alluin, J., and Rossi, J.J. (2007). Distance constraints between microRNA target sites dictate efficacy and cooperativity. Nucleic Acids Res *35*, 2333–2342.

Saito, Y., and Jones, P. (2006). Epigenetic Activation of Tumor Suppressor MicroRNAs in Human Cancer Cells. Cell Cycle *5*, 2220–2222.

Saj, A., and Lai, E.C. (2011). Control of microRNA biogenesis and transcription by cell signaling pathways. Curr. Opin. Genet. Dev. *21*, 504–510.

Sales, G., Coppe, A., Bisognin, A., Biasiolo, M., Bortoluzzi, S., and Romualdi, C. (2010). MAGIA, a web-based tool for miRNA and Genes Integrated Analysis. Nucleic Acids Res *38*, W352–W359.

Salwinski, L., and Eisenberg, D. (2003). Computational methods of analysis of protein–protein interactions. Curr. Opin. Struct. Biol. *13*, 377–382.

Santra, T., Kolch, W., and Kholodenko, B.N. (2014). Navigating the Multilayered Organization of Eukaryotic Signaling: A New Trend in Data Integration. PLoS Comput. Biol. *10*, e1003385.

Savageau, M.A., Coelho, P.M.B.M., Fasani, R.A., Tolla, D.A., and Salvador, A. (2009). Phenotypes and tolerances in the design space of biochemical systems. Proc. Natl. Acad. Sci. U. S. A. *106*, 6435–6440.

Sax, J.K., and El-Deiry, W.S. (2003). p53 downstream targets and chemosensitivity. Cell Death Differ. *10*, 413–417.

Schmidt, H., and Jirstrand, M. (2006). Systems Biology Toolbox for MATLAB: a computational platform for research in systems biology. Bioinformatics *22*, 514–515.

Schmitter, D., Filkowski, J., Sewer, A., Pillai, R.S., Oakeley, E.J., Zavolan, M., Svoboda, P., and Filipowicz, W. (2006). Effects of Dicer and Argonaute down-regulation on mRNA levels in human HEK293 cells. Nucleic Acids Res. *34*, 4801–4815.

Schmitz, U. (2013). MicroRNA target regulation. In Encyclopedia of Systems Biology, W. Dubitzky, O. Wolkenhauer, K.-H. Cho, and H. Yokota, eds. (Springer New York), pp. 1346–1350.

Schmitz, U., and Vearasilp, K. (2013). miRBase. In Encyclopedia of Systems Biology, W. Dubitzky, O. Wolkenhauer, K.-H. Cho, and H. Yokota, eds. (Springer New York), pp. 1363–1366.

Schmitz, U., and Wolkenhauer, O. (2013). Web Resources for microRNA Research. In MicroRNA Cancer Regulation, U. Schmitz, O. Wolkenhauer, and J. Vera, eds. (Dordrecht: Springer Netherlands), pp. 225–250.

Schmitz, U., Wolkenhauer, O., and Vera, J. (2013b). MicroRNA Cancer Regulation Advanced Concepts, Bioinformatics and Systems Biology Tools. (Dordrecht: Springer).

Schmitz, U., Lai, X., Winter, F., Wolkenhauer, O., Vera, J., and Gupta, S.K. (2014). Cooperative gene regulation by microRNA pairs and their identification using a computational workflow. Nucleic Acids Res. *42*, 7539–7552.

Schultz, J., Ibrahim, S.M., Vera, J., and Kunz, M. (2009). 14-3-3sigma gene silencing during melanoma progression and its role in cell cycle control and cellular senescence. Mol. Cancer *8*, 53.

Schwab, R., Palatnik, J.F., Riester, M., Schommer, C., Schmid, M., and Weigel, D. (2005). Specific Effects of MicroRNAs on the Plant Transcriptome. Dev. Cell *8*, 517–527.

Schwarzenbach, H., Hoon, D.S.B., and Pantel, K. (2011). Cell-free nucleic acids as biomarkers in cancer patients. Nat Rev Cancer *11*, 426–437.

Selbach, M., Schwanhäusser, B., Thierfelder, N., Fang, Z., Khanin, R., and Rajewsky, N. (2008). Widespread changes in protein synthesis induced by microRNAs. Nature *455*, 58–63.

Sethupathy, P., Megraw, M., and Hatzigeorgiou, A.G. (2006). A guide through present computational approaches for the identification of mammalian microRNA targets. Nat. Methods *3*, 881–886.

Sethupathy, P., Corda, B., and Hatzigeorgiou, A.G. (2006d). TarBase: A comprehensive database of experimentally supported animal microRNA targets. RNA *12*, 192–197.

Seto, A.G. (2010). The road toward microRNA therapeutics. Int. J. Biochem. Cell Biol. *42*, 1298–1305.

Shalgi, R., Lieber, D., Oren, M., and Pilpel, Y. (2007). Global and local architecture of the mammalian microRNA-transcription factor regulatory network. PLoS Comput Biol *3*, e131.

Shannon, P. (2003). Cytoscape: A Software Environment for Integrated Models of Biomolecular Interaction Networks. Genome Res. *13*, 2498–2504.

Sharan, R., and Ideker, T. (2006). Modeling cellular machinery through biological network comparison. Nat. Biotechnol. *24*, 427–433.

Shen, L., Johnson, T.L., Clugston, S., Huang, H., Butenhof, K.J., and Stanton, R.V. (2011). Molecular Dynamics Simulation and Binding Energy Calculation for Estimation of Oligonucleotide Duplex Thermostability in RNA-Based Therapeutics. J. Chem. Inf. Model. *51*, 1957–1965.

Siepel, A., Bejerano, G., Pedersen, J.S., Hinrichs, A.S., Hou, M., Rosenbloom, K., Clawson, H., Spieth, J., Hillier, L.W., Richards, S., *et al.* (2005). Evolutionarily conserved elements in vertebrate, insect, worm, and yeast genomes. Genome Res. *15*, 1034–1050.

Sinha, A.U., Kaimal, V., Chen, J., and Jegga, A.G. (2008). Dissecting microregulation of a master regulatory network. BMC Genomics *9*, 88.

Soifer, H.S., Rossi, J.J., and Sætrom, P. (2007). MicroRNAs in Disease and Potential Therapeutic Applications. Mol. Ther. *15*, 2070–2079.

Stark, A., Brennecke, J., Russell, R.B., and Cohen, S.M. (2003). Identification of Drosophila MicroRNA targets. PLoS Biol *1*, E60.

Subramanian, A. (2005). From the Cover: Gene set enrichment analysis: A knowledge-based approach for interpreting genome-wide expression profiles. Proc. Natl. Acad. Sci. *102*, 15545–15550.

Szklarczyk, D., Franceschini, A., Kuhn, M., Simonovic, M., Roth, A., Minguez, P., Doerks, T., Stark, M., Muller, J., Bork, P., *et al.* (2011). The STRING database in 2011: functional interaction networks of proteins, globally integrated and scored. Nucleic Acids Res. *39*, D561–D568.

Tétreault, N., and De Guire, V. (2013). miRNAs: Their discovery, biogenesis and mechanism of action. Clin. Biochem. *46*, 842–845.

Thadani, R., and Tammi, M. (2006). MicroTar: predicting microRNA targets from RNA duplexes. BMC Bioinformatics *7*, S20.

Thum, T., Gross, C., Fiedler, J., Fischer, T., Kissler, S., Bussen, M., Galuppo, P., Just, S., Rottbauer, W., Frantz, S., *et al.* (2008). MicroRNA-21 contributes to myocardial disease by stimulating MAP kinase signalling in fibroblasts. Nature *456*, 980–984.

Tsang, J., Zhu, J., and van Oudenaarden, A. (2007). MicroRNA-Mediated Feedback and Feedforward Loops Are Recurrent Network Motifs in Mammals. Mol. Cell *26*, 753–767.

Tu, K., Yu, H., Hua, Y.-J., Li, Y.-Y., Liu, L., Xie, L., and Li, Y.-X. (2009). Combinatorial network of primary and secondary microRNA-driven regulatory mechanisms. Nucleic Acids Res *37*, 5969–5980.

Ueda, T., Volinia, S., Okumura, H., Shimizu, M., Taccioli, C., Rossi, S., Alder, H., Liu, C., Oue, N., Yasui, W., *et al.* (2010). Relation between microRNA expression and progression and prognosis of gastric cancer: a microRNA expression analysis. Lancet Oncol. *11*, 136–146.

Ulitsky, I., Laurent, L.C., and Shamir, R. (2010). Towards computational prediction of microRNA function and activity. Nucleic Acids Res *38*, e160.

Vella, M.C., Choi, E.-Y., Lin, S.-Y., Reinert, K., and Slack, F.J. (2004). The C. elegans microRNA let-7 binds to imperfect let-7 complementary sites from the lin-41 3'UTR. Genes Dev. *18*, 132–137.

Vera, J., Rath, O., Balsa-Canto, E., Banga, J.R., Kolch, W., and Wolkenhauer, O. (2010). Investigating dynamics of inhibitory and feedback loops in ERK signalling using power-law models. Mol. Biosyst. *6*, 2174.

Vera, J., Lai, X., Schmitz, U., and Wolkenhauer, O. (2013b). MicroRNA-Regulated Networks: The Perfect Storm for Classical Molecular Biology, the Ideal Scenario for Systems Biology. In MicroRNA Cancer Regulation, U. Schmitz, O. Wolkenhauer, and J. Vera, eds. (Dordrecht: Springer Netherlands), pp. 55–76.

Vera, J., Schmitz, U., Lai, X., Engelmann, D., Khan, F.M., Wolkenhauer, O., and Putzer, B.M. (2013c). Kinetic Modeling-Based Detection of Genetic Signatures that Provide Chemoresistance via the E2F1-p73/DNp73-miR-205 Network. Cancer Res. *73*, 3511–3524.

Vera, J., Wolkenhauer, O., and Schmitz, U. (2014). Current Achievements in Cancer Systems Biology. In eLS., (John Wiley & Sons Ltd, Chichester. http://www.els.net),.

Vergoulis, T., Vlachos, I.S., Alexiou, P., Georgakilas, G., Maragkakis, M., Reczko, M., Gerangelos, S., Koziris, N., Dalamagas, T., and Hatzigeorgiou, A.G. (2011). TarBase 6.0:

capturing the exponential growth of miRNA targets with experimental support. Nucleic Acids Res. *40*, D222–D229.

Vitale, I., Galluzzi, L., Castedo, M., and Kroemer, G. (2011). Mitotic catastrophe: a mechanism for avoiding genomic instability. Nat. Rev. Mol. Cell Biol. *12*, 385–392.

Vohradsky, J., Panek, J., and Vomastek, T. (2010). Numerical modelling of microRNA-mediated mRNA decay identifies novel mechanism of microRNA controlled mRNA downregulation. Nucleic Acids Res *38*, 4579–4585.

Volinia, S., Visone, R., Galasso, M., Rossi, E., and Croce, C.M. (2009). Identification of microRNA activity by Targets' Reverse EXpression. Bioinformatics *26*, 91–97.

Wain, H.M., Bruford, E.A., Lovering, R.C., Lush, M.J., Wright, M.W., and Povey, S. (2002). Guidelines for human gene nomenclature. Genomics *79*, 464–470.

Wang, X., and El Naqa, I.M. (2007). Prediction of both conserved and nonconserved microRNA targets in animals. Bioinformatics *24*, 325–332.

Wang, X., and Wang, X. (2006). Systematic identification of microRNA functions by combining target prediction and expression profiling. Nucleic Acids Res *34*, 1646–1652.

Wang, J., Lu, M., Qiu, C., and Cui, Q. (2010). TransmiR: a transcription factor-microRNA regulation database. Nucleic Acids Res. *38*, D119–D122.

Wang, W., Furneaux, H., Cheng, H., Caldwell, M.C., Hutter, D., Liu, Y., Holbrook, N., and Gorospe, M. (2000). HuR regulates p21 mRNA stabilization by UV light. Mol. Cell. Biol. *20*, 760–769.

Wang, Y., Juranek, S., Li, H., Sheng, G., Tuschl, T., and Patel, D.J. (2008). Structure of an argonaute silencing complex with a seed-containing guide DNA and target RNA duplex. Nature *456*, 921–926.

Washietl, S., Hofacker, I.L., and Stadler, P.F. (2005). Fast and reliable prediction of noncoding RNAs. Proc Natl Acad Sci U A *102*, 2454–2459.

Willingham, A.T. (2005). A Strategy for Probing the Function of Noncoding RNAs Finds a Repressor of NFAT. Science *309*, 1570–1573.

Winter, J., Jung, S., Keller, S., Gregory, R.I., and Diederichs, S. (2009). Many roads to maturity: microRNA biogenesis pathways and their regulation. Nat Cell Biol *11*, 228–234.

Wu, D., Rice, C.M., and Wang, X. (2012). Cancer bioinformatics: A new approach to systems clinical medicine. BMC Bioinformatics *13*, 71.

Wu, S., Huang, S., Ding, J., Zhao, Y., Liang, L., Liu, T., Zhan, R., and He, X. (2010). Multiple microRNAs modulate p21Cip1/Waf1 expression by directly targeting its 3' untranslated region. Oncogene *29*, 2302–2308.

Xiao, F., Zuo, Z., Cai, G., Kang, S., Gao, X., and Li, T. (2009). miRecords: an integrated resource for microRNA-target interactions. Nucleic Acids Res. *37*, D105–D110.

Xie, Z.-R., Yang, H.-T., Liu, W.-C., and Hwang, M.-J. (2007). The role of microRNA in the delayed negative feedback regulation of gene expression. Biochem Biophys Res Commun *358*, 722–726.

Xu, F., Liu, Z., Shen, J., and Wang, R. (2009). Dynamics of microRNA-mediated motifs. IET Syst Biol *3*, 496–504.

Yan, X., Chao, T., Tu, K., Zhang, Y., Xie, L., Gong, Y., Yuan, J., Qiang, B., and Peng, X. (2007). Improving the prediction of human microRNA target genes by using ensemble algorithm. FEBS Lett *581*, 1587–1593.

Yekta, S. (2004). MicroRNA-Directed Cleavage of HOXB8 mRNA. Science *304*, 594–596.

Zadeh, J.N., Steenberg, C.D., Bois, J.S., Wolfe, B.R., Pierce, M.B., Khan, A.R., Dirks, R.M., and Pierce, N.A. (2011). NUPACK: Analysis and design of nucleic acid systems. J. Comput. Chem. *32*, 170–173.

Zampetaki, A., Willeit, P., Drozdov, I., Kiechl, S., and Mayr, M. (2011). Profiling of circulating microRNAs: from single biomarkers to re-wired networks. Cardiovasc. Res. *93*, 555–562.

Zhao, H., Shen, J., Medico, L., Wang, D., Ambrosone, C.B., and Liu, S. (2010). A pilot study of circulating miRNAs as potential biomarkers of early stage breast cancer. PLoS One *5*, e13735.

Zhao, H., Yang, Y., and Zhou, Y. (2013a). Prediction of RNA binding proteins comes of age from low resolution to high resolution. Mol. Biosyst. *9*, 2417.

Zhao, M., Sun, J., and Zhao, Z. (2013b). Synergetic regulatory networks mediated by oncogene-driven microRNAs and transcription factors in serous ovarian cancer. Mol. Biosyst. *9*, 3187–3198.

Zinovyev, A., Morozova, N., Nonne, N., Barillot, E., Harel-Bellan, A., and Gorban, A.N. (2010). Dynamical modeling of microRNA action on the protein translation process. BMC Syst. Biol. *4*, 13.

Zinovyev, A., Morozova, N., Gorban, A.N., and Harel-Belan, A. (2013). Mathematical Modeling of microRNA–Mediated Mechanisms of Translation Repression. In MicroRNA Cancer Regulation, U. Schmitz, O. Wolkenhauer, and J. Vera, eds. (Dordrecht: Springer Netherlands), pp. 189–224.

Zuker, M. (2003). Mfold web server for nucleic acid folding and hybridization prediction. Nucleic Acids Res. *31*, 3406–3415.

Curriculum vitae

PERSONAL DETAILS

Name	Ulf Schmitz
Address	Gutshof 11 18059 Papendorf, Germany
Date of birth	8 July 1978, Wismar
Phone	+49 381 498 7576
Email	ulf.schmitz@uni-rostock.de
Web	http://www.sbi.uni-rostock.de/

EMPLOYMENT

2003 to date

Bioinformatician and Systems Engineer
Department of Systems Biology & Bioinformatics, University of Rostock

Responsibilities:

- teaching assistance
- supervision of student projects
- software and hardware maintenance
- software development (Python, Perl, Matlab)
- web programming
- organisation of workshops/conferences
- supervision of apprentices
- support in grant application writing

03/2002 – 08/2002

Internship at Pass IT Consulting P LTD in Hyderabad, India
"Development work related Enterprise Software for the Road Transport Corporation of the State of Andhra Pradesh (APSRTC)"

02/2001 – 07/2001

Internship at D&S Automation GmbH in Schwerin, Germany:
"Development of modules for an ANSI-C based real-time MJPEG-Encoder/Decoder"

EDUCATION

2013 to date

PhD in Bioinformatics, Faculty of Computer Science and Electrical Engineering, University of Rostock

1998 - 2003

Dipl.-Ing. (FH) in Multimedia Engineering at the University of Applied Sciences Wismar, Specialization on Medical Informatics

Final year project: *Development of software for testing hearing abilities of humans*

1991 - 1997

Abitur (A-level equivalent) at Helene-Weigel-Gymnasium, Wismar

TRAINING

01/2014	Workshop: 'Clinical needs in oncology and cardiovascular diseases as drivers for a Systems Medicine Approach', Genoa, Italy
03/2014	Workshop: How to write a winning proposal in Horizon 2020
12/2013	Workshop on: Systems immunology, University of Erlangen-Nuremberg
10/2013	Workshop for training presentation skills and journalistic writing
09/2012	International workshop on small RNA in cancer, inflammation and aging
09/2010	Workshop on EBI resources and data integration (EBI Roadshow)
06/2009	Transatlantic Summer School on Cancer Systems Biology (CaSysBio), Rostock/Warnemünde, Germany
10/2008	Systems Biology for Medical Applications Summer School, Costa Adeje, Tenerife, Spain
03/2007	Winter School on Systems Biology for Medical Applications, Puerto de La Cruz, Tenerife, Spain
04/2007	Tage der Lehre - professional training at the University of Rostock on: • Organizing workshops and seminars • Presentation skills • Rhetoric – Training with theatre methods
03/2002 – 08/2002	Business Information Technology (one semester) at the Centre for Management and Information Technology (CMIT) in Hyderabad/India
04/2002 – 07/2002	MCSE Windows 2000 course at the Gatestech Education & Training Division in Hyderabad / India

REVIEWS FOR JOURNALS/ADVISORY FUNCTIONS

- Member of the Program Committee of the IEEE International Conference on Bioinformatics and Biomedicine (BIBM, 2012 - to date)
- Reviewer for the journals: Bioinformatics, BMC Bioinformatics, BMC Systems Biology, Databases, Nucleic Acids Research, Fungal Genetics and Biology, PLOS ONE, Current Cancer Drug Targets
- Reviewer for Encyclopaedia Römpp Online

CONTRIBUTIONS TO GRANT PROPOSALS

2014	MelEVIR - Melanoma, Extracellular Vesicles and Immune Response. BMBF (e:Bio II).
2014	OASYS - Targeting OsteoArthritis; a SYStems medicine approach. EU H2020 - Understanding disease; systems medicine (PHC2)
2014	PRISM - Understanding Brain Damage in Preterm Infants Using Systems Biomedicine. EU H2020 - Understanding disease; systems medicine (PHC2)
2014	CaSyME - Understanding Cancer through Systems Medicine approaches in the Elderly: from mathematical modelling to the clinic. EU H2020 - Understanding disease; systems medicine (PHC2)

2013	Mechanistic approaches toward cardiac regeneration. SFB Vorantrag
2013	A systems medicine approach to identify circulating microRNAs as biomarker for neonatal septicaemia. DLR - Richtlinien zur Förderung der Wissenschaftlich-Technologischen Zusammenarbeit (WTZ) mit Indien/ ICMR (SepticmiR)
2013	MelamiRSys - miRNAs and the interaction between melanoma and the immune system. A systems biology approach for developing a melanoma miRNA blood test. BMBF (Sysmed2) - Demonstratoren zur Individualisierten Medizin
2013	ROCSYM - Rostock Center for Systems Medicine – Cardiovascular Regeneration. BMBF (e:Med)
2012	Rolle E2F1 regulierter microRNAs bei der Tumourmetastasierung und deren antimetastatisches Therapiepotential, DFG
2011	miRSys – microRNA at the crossroad of lung inflammation, regeneration and cancer. BMBF
2011	Multiple Organ Failure (MOF): From Systems Biology towards Clinical Utility (eBio), BMBF
2009	Identification and functional analysis of microRNAs involved in malignant melanoma progression. 2009-2011, DFG Sachbeihilfe (WO 991/4-1)

TEACHING and SUPERVISION

2014	Supervision of MSc project (Markus Wolfien): Next Generation Sequencing Data Analysis of Stem Cell Derived Cardiomyocyte Cell Types. In collaboration with Prof. Robert David (University Medicine Rostock)
2013	Supervision of MSc project (Nomad Al Hassan): Identification and analysis of putatively cooperating microRNAs.
2013	Supervision of MSc project (Mustafa Baig Mirza): Extending the multi-valued logic approach for modelling biochemical reaction networks.
2012	Supervision of MSc project (Afi Akoele Blitti): Computational identification of ADAR editing sites in human microRNA precursors.
2008	Supervision of Diplom project (Peter Raasch): A Java framework to integrate ncRNA detection methods (published in BMC Bioinformatics, see Raasch *et al.* 2010)
2006 - 2011	Three collaborative software development projects (between 3-10 students, 4 SWS)
2005 - 2009	Teaching assistant 'Data Handling Skills' (4 SWS)
2004	Teaching assistant 'Bioinformatics programming languages' (Perl/Python; 2 SWS)

OTHER ACTIVITIES

- Supervision of pupils (in total 5) from local schools completing a placement at the Department of Systems Biology & Bioinformatics

- Co-organizer of international conferences:
  - Transatlantic Summer School on Cancer Systems Biology (CaSysBio), 06/2009, Rostock, Germany
  - Systems Biology for Medical Applications Summer School, 10/2008, Costa Adeje, Tenerife, Spain
  - Winter School on Systems Biology for Medical Applications, 03/2007, Puerto de La Cruz, Tenerife, Spain

HONORS + AWARDS

2014	Dechema funded travel grant for attending the GCB 2014, Bielefeld
2014	Award for the "Model of the Month" in the BioModels database
2014	BMBF funded travel grant for attending the SBHD 2014, Boston, USA
2013	Faculty award for extraordinary work performance
2013	Dechema funded travel grant for attending the GCB 2014, Göttingen
2009	Award for the best scientific contribution to the public image of the Faculty of Computer Science and Electrical Engineering, University of Rostock

Publications

Journal articles (peer reviewed)

A Bhattacharya, **U Schmitz**, Y Raatz, M Schönherr, T Kottek, M Schauer, S Franz, A Saalbach, U Anderegg, O Wolkenhauer, D Schadendorf, JC Simon, T Magin, J Vera, M Kunz (2015) miR-638 promotes melanoma metastasis and protects melanoma cells from apoptosis and autophagy. Oncotarget, accepted for publication.

R Amirkhah, **U Schmitz**, M Linnebacher, O Wolkenhauer, A Farazmand (2014). MicroRNA-mRNA interactions in colorectal cancer and their role in tumor progression. *Genes, Chromosomes and Cancer,* accepted for publication.

S Knoll, K Fürst, B Kowtharapu, **U Schmitz**, S Marquardt, O Wolkenhauer, H Martin, BM Pützer (2014) MiR-224/452 induction through E2F1 drives epithelial-mesenchymal transition by targeting TXNIP. *EMBO Reports*, 15, 1315–1329.

U Schmitz, X Lai, F Winter, O Wolkenhauer, J Vera, S Gupta (2014) Cooperative gene regulation by microRNA pairs and their identification using a computational workflow. *Nucleic Acid Research*, 42, 12, p. 7539-7552.

FM Khan, **U Schmitz**, S Nikolov, D Engelmann, BM Pützer, O Wolkenhauer, J Vera (2014) Hybrid modelling of the crosstalk between signaling and transcriptional networks using ordinary differential equations and multi-valued logic. *Biochim Biophys Acta*, 1844 (1 Pt B): 289-98.

X Lai, A Bhattacharya, **U Schmitz**, M Kunz, J Vera, O Wolkenhauer (2013) A systems biology approach to study microRNA-mediated gene regulatory networks. *BioMed Research International*, vol. 2013:703849.

J Vera, **U Schmitz**, X Lai, D Engelmann, FM Khan, O Wolkenhauer, BM Pützer (2013) Kinetic modelling-based detection of genetic signatures that provide chemoresistance via the E2F1-p73/DNp73-miR-205 network. *Cancer Research*, 73(12):3511-2.

A Bhattacharaya, **U Schmitz**, O Wolkenhauer, M Schönherr, Y Raatz, M Kunz (2013) Regulation of cell cycle checkpoint kinase Wee1 by miR-195 in malignant melanoma. *Oncogene*, 32, 3175–3183.

G Fuellen, J Dengjel, A Hoeflich, J Hoeijmakers, H Kestler, A Kowald, S Priebe, D Rebholz-Schuhmann, B Schmeck, **U Schmítz**, A Stolzing, J Sühnel, Fritz Lipmann, D Wuttke, J Vera (2012) Systems Biology and Bioinformatics in Aging Research: A Workshop Report. *Rejuvenation Research*, 15(6): 631-641.

V Alla, BS Kowtharapu, D Engelmann, S Emmrich, **U Schmitz**, M Steder, BM Pützer (2012) E2F1 confers anticancer drug resistance by targeting ABC transporter family members and Bcl-2 via the p73/DNp73-miR-205 circuitry. *Cell Cycle*, 11:1.

X Lai*, **U Schmitz***, S Gupta, A Bhattacharya, M Kunz, O Wolkenhauer, J Vera (2012) Computational analysis of target hub gene repression regulated by multiple and cooperative miRNAs. *Nucleic Acid Research*, 40: 8818-8834, (* eq. contr. authors).

D Guebel*, **U Schmitz***, O Wolkenhauer, J Vera (2012) Analysis of cell adhesion in stem cells during early stages of colon cancer based on an extended multi-valued logic approach. *Mol.Biosyst.*, DOI:10.1039/c2mb05277f, (* eq. contr. authors).

SK Gupta, SK Gupta, S Smita, M Srivastava, X Lai, **U Schmitz**, Q Rahman, O Wolkenhauer, J Vera (2011) Computational analysis and modelling the effectiveness of 'Zanamivir' targeting neuraminidase protein in pandemic H1N1 strains. *Infection, Genetics and Evolution*, Volume 11, Issue 5, pp. 1072-1082.

SK Gupta, M Srivastava, BA Akhoon, S Smita, **U Schmitz**, O Wolkenhauer, J Vera, SK Gupta (2011) Identification of immunogenic consensus T-cell epitopes in globally distributed influenza-A H1N1 neuraminidase. *Infection, Genetics and Evolution,* (2):308-19.

S Nikolov, J Vera, **U Schmitz**, O Wolkenhauer (2011) A model-based strategy to investigate the role of microRNA regulation in cancer signalling networks. *Theory in Biosciences,* Volume 130, Issue 1, pp 55-69.

P Raasch, **U Schmitz**, N Patenge, J Vera, B Kreikemeyer, O Wolkenhauer (2010) non-coding RNA detection methods combined to improve usability, reproducibility and precision. *BMC Bioinformatics,* 11:491.

J Schultz, D Koczan, **U Schmitz**, SM Ibrahim, D Pilch, J Landsberg, M Kunz (2010) Tumour-promoting role of signal transducer and activator of transcription (Stat)1 in late-stage melanoma growth. *Clin Exp Metastasis,* 27(3):133-40.

J Vera, T Kwon, **U Schmitz**, W Kolch, O Wolkenhauer (2009) Exploration of homodimer receptor - homodimer protein interactions. *International Journal of Bioinformatics Research and Applications,* Vol. 5, No.4 pp. 447-457.

R Hoffrogge, S Beyer, R Hübner, S Mikkat, E Mix, C Scharf, **U Schmitz**, S Pauleweit, M Berth, IZ Zbrzycki, H Christoph, J Pahnke, O Wolkenhauer, A Uhrmacher, U Völker, R Rolfs (2007) 2-DE profiling of GDNF overexpression-related proteome changes in differentiating ST14A rat progenitor cells. *Proteomics*, 7, 33-46.

Journal articles (submitted)

M Schauer, T Kottek, M Schönherr, A Bhattacharya, JC Simon, SM Ibrahim, M Hirose, M Möller, M Mörl, R Köhling, G Füllen, **U Schmitz**, M Kunz (2014) A mutation in the NADH-dehydrogenase subunit 2 negatively interferes with fibroblast ageing. Submitted for publication.

Books and Chapters

J Vera, O Wolkenhauer, **U Schmitz** (2014) Cancer Systems Biology: Current Achievements in – Omics Data Analysis, Network Reconstruction and Mathematical Modelling. In: eLS. John Wiley & Sons Ltd, Chichester. www.els.net.

U Schmitz, O Wolkenhauer, J Vera (eds.). MicroRNA cancer regulation: Advanced concepts, bioinformatics and systems biology tools. Series: Advances in Experimental Medicine and Biology, Vol. 774, Springer 2013. ISBN: 978-94-007-5589-5.

U Schmitz, O Wolkenhauer (2013) Web resources for microRNA research. In: MicroRNA Cancer Regulation: Advance Concepts, Bioinformatics and Systems Biology Tools. Series: Advances in Experimental Medicine and Biology, Vol. 774, pp. 225-250.

J Vera, X Lai, **U Schmitz**, O Wolkenhauer (2013) MicroRNA-regulated networks: the perfect storm for classical molecular biology, the ideal scenario for systems biology. In: MicroRNA Cancer Regulation: Advance Concepts, Bioinformatics and Systems Biology Tools. Series: Advances in Experimental Medicine and Biology, Vol. 774, pp. 55–76.

U Schmitz, J Vera (2013) Computational microRNA biology. In Encyclopedia of Systems Biology. W Dubitzky, O Wolkenhauer, H Yokota, Q Cho (eds. in chief), Springer New York, ISBN-13: 9781441998620.

S Gupta, **U Schmitz** (2011) Bioinformatics Analysis of High-Throughput Experiments. In Recent Trends in Biotechnology. Volume 2, MP Singh, A Agrawal, B Sharma (eds.), *Nova Science Publishers.*

Published Reports

P Raasch, **U Schmitz**, B Kreikemeyer, O Wolkenhauer (2009) Effective ncRNA detection. *Rostocker Informatik-Berichte*, 32.

J Vera, T Kwon, **U Schmitz**, T Millat, O Wolkenhauer (2007) Bioinformatics analysis of interactions of receptor homodimers with intracellular protein homodimers. *Rostocker Informatik-Berichte,* 31.

Encyclopedia entries

S Freiesleben, **U Schmitz**, J Vera (2013) Regulation of miRNA biogenesis. In: *Encyclopaedia of Systems Biology*, W Dubitzky, O Wolkenhauer, H Yokata, K-H Cho (eds.), Springer New York.

K Vearasilp, **U Schmitz** (2013) microRNA web resources. In: *Encyclopaedia of Systems Biology*. W Dubitzky, O Wolkenhauer, H Yokata, K-H Cho (eds.), Springer New York.

U Schmitz, K Vearasilp (2013) miRBase. In: *Encyclopaedia of Systems Biology*. W Dubitzky, O Wolkenhauer, H Yokata, K-H Cho (eds.), Springer New York.

U Schmitz, SGupta (2013) Target site. In: *Encyclopaedia of Systems Biology*. W Dubitzky, O Wolkenhauer, H Yokata, K-H Cho (eds.), Springer New York.

U Schmitz (2013) Target regulation. In: *Encyclopaedia of Systems Biology*. W Dubitzky, O Wolkenhauer, H Yokata, K-H Cho (eds.), Springer New York.

DV Guebel, **U Schmitz**, J Vera (2013) miRNA embedding regulation networks, logical modelling. In: *Encyclopaedia of Systems Biology*. W Dubitzky, O Wolkenhauer, H Yokata, K-H Cho (eds.), Springer New York.

Conference Contributions (Posters)

SK Gupta, **U Schmitz**, X Lai, J Vera, O Wolkenhauer: Cooperative microRNA regulation in anti-cancer drug resistance - a computational approach. German Conference on Bioinformatics (GCB 2014), Bielefeld, 2014.

U Schmitz, SK Gupta, F Winter, O Wolkenhauer: Cooperative gene regulation by microRNA pairs and their identification through an integrative workflow. International Conference on Systems Biology of Human Diseases (SBHD 2014), Boston, USA, 2014.

A Bhattacharya, **U Schmitz**, Y Raatz, M Schonherr, T Kottek, M Schauer, *et al.* miR-638 promotes melanoma metastasis and is repressed by transcription factor AP2-alpha. Journal of Investigative Dermatology, 134, S128-S128, 2014.

U Schmitz, SK Gupta, X Lai, J Vera, O Wolkenhauer. Predicting targets of synergistic microRNA regulation. German Conference on Bioinformatic, Göttingen, 2013.

A Bhattacharya, **U Schmitz**, Y Raatz, M Schoenherr, M Kunz. New candidate oncomIRs for malignant melanoma. Journal of Investigative Dermatology, 133, S230-S230, 2013.

U Schmitz, X Lai, J Vera, O Wolkenhauer. The regulation of microRNA target hubs. International Workshop on Small RNA in Cancer, Inflammation, and Aging, Copenhagen (Denmark), September 3 - 4 , 2012.

X Lai, **U Schmitz**, J Vera, O Wolkenhauer. A systems biology approach to study the cellular function of microRNAs. International Workshop on Small RNA in Cancer, Inflammation, and Aging, Copenhagen (Denmark), September 3 - 4, 2012.

X Lai, **U Schmitz**, M Kunz, O Wolkenhauer, J Vera: On the regulation of microRNA target hubs: a Systems Biology perspective. 12th International Conference on Systems Biology (ICSB2010). Heidelberg, Germany, August 2011.

A Bhattacharya, **U Schmitz**, M Schonherr, Y Raatz, M Kunz: Role of microRNA 372 in melanoma progression. Experimental Dermatology, Vol 20, Issue 2, 196, 2011.

DV Guebel, **U Schmitz**, O Wolkenhauer, J Vera. Análisis de los fenómenos de adhesión celular en los estadios tempranos de cáncer de colon mediante un modelo de lógica multi-valuada extendida. "Angel H. Roffo" (Facultad de Medicina, Universidad de Buenos Aires). Buenos Aires, Argentina. 13-16 Sept. 2011.

A Bhattacharaya, **U Schmitz**, M Schonherr, Y Raatz, M Kunz. Role of micro RNA 372 in melanoma progression. *Experimental Dermatology*, Volume 20, Issue 2, 196, 2011

D Guebel, **U Schmitz**, O Wolkenhauer, J Vera. An extensive Boolean network for analysing the signalling events in early stages of humen colon cancer. *11th International Conference on Systems Biology (ICSB)*, Edinburgh, October 2010.

U Schmitz, A Bhattacharya, S Gupta, J Schultz, M Kunz, O Wolkenhauer. MicroRNAs in malignant melanoma progression. *German Conference on Bioinformatics*, Braunschweig, 2010.

P Raasch, **U Schmitz**, N Patenge, J Vera, B Kreikemeyer and O Wolkenhauer. Local Genomic Properties as Indicators of Functional Elements. *3. Gemeinsame Tagung der DGHM und der VAAM*, 2010, Hannover, Germany.

P Raasch, **U Schmitz**, B Kreikemeyer, O Wolkenhauer. A JAVA framework to integrate ncRNA detection methods. *Systems Biology for Medical Applications Summer School*, Costa Adeje, Spain, 2008.

E Stalidzans, I Mozga, Ch Roos, E Fernandez, J Font, J Vanags, M Remm, **U Schmitz**, O Wolkenhauer. Modular education for interdisciplinary Systems Biology. *9th International Conference on Systems Biology*, Gothenburg, 2008.

J Vera, T Millat, T Kwon, **U Schmitz**, O Wolkenhauer. Exploring cell signalling homodimer receptor-homodimer protein interactions and their dynamical consequences. *German Conference on Bioinformatics*, Potsdam, 2007.

Scientific Talks

Cooperative microRNA regulation in anti-cancer drug resistance - a computational approach. Poster Flash Presentation, German Conference on Bioinformatics (GCB 2014), Bielefeld, October 2014.

Cooperative gene regulation by microRNA pairs - identification and analysis in a computational workflow. Wissensforum, Universitätsmedizin Rostock, April 2014.

Computergestützte Genregulierung von der Petrischale bis zum Krankenbett. Tag der Promovierenden, Universität Rostock, April 2014.

Computational microRNA biology. Rostock Symposium of Systems Biology in Aging (ROSYBA), Rostock, 2011.

Genomics and Sequence Analysis. MOSBIO - International introductory course in systems biology, Espoo, Finland, June 2009.

Systems Biology - Software, Data Exchange Standards and Ontologies. MOSBIO - International introductory course in systems biology, Bilbao, Spain, May 2009.

Theses

1. The complexity of networks regulating miRNA target hubs necessitates an integrative approach combining bioinformatics and systems biology methods (Chapter 3).

2. Through the development of workflows, it is possible to characterize miRNA involved regulatory networks as well as predict and analyse RNA triplexes (Chapters 3 & 4).

3. The integration of data on molecular interactions reveals sophisticated linkages among miRNAs and TFs for the fine-tuned regulation of miRNA target hub genes (Section 3.4.1).

4. Model-based simulations reveal landscapes of target gene repression. These can be used to test consequences of different modes of combinatorial target regulation by miRNAs (Section 3.4.3).

5. Mathematical modelling predicts target gene expression under concerted and/or cooperative miRNA-induced repression, even for different cell functions (Section 3.4.5).

6. Modulating the strength of miRNA cooperativity influences target repression efficiency and can be used to buffer transcriptional noise (Section 3.4.7).

7. *In silico* analyses generate the hypothesis that cooperative target regulation effects most human protein coding genes (Section 4.3.2).

8. Secondary structure predictions can be used to determine the thermodynamically most stable RNA triplex structures (Section 4.3.4).

9. Molecular dynamics simulations can be used to identify thermodynamically unstable RNA triplexes (Section 4.3.5).

10. Gene set enrichment analysis reveals the relevance of miRNA functions for disease phenotypes (Section 4.4.5).

11. The workflows described in this work can be applied in scenarios where the role of miRNA regulation in the control of single genes, (signalling) pathways and larger networks driving cellular processes is investigated (Chapter 5).